Advances in Industrial Control

Springer
*London
Berlin
Heidelberg
New York
Barcelona
Hong Kong
Milan
Paris
Santa Clara
Singapore
Tokyo*

Other titles published in this Series:

System Identification and Robust Control
Steen Tøffner-Clausen

Genetic Algorithms for Control and Signal Processing
K.F. Man, K.S. Tang, S. Kwong and W.A. Halang

Advanced Control of Solar Plants
E.F. Camacho, M. Berenguel and F.R. Rubio

Control of Modern Integrated Power Systems
E. Mariani and S.S. Murthy

Advanced Load Dispatch for Power Systems: Principles, Practices and Economies
E. Mariani and S.S. Murthy

Supervision and Control for Industrial Processes
Björn Sohlberg

Modelling and Simulation of Human Behaviour in System Control
Pietro Carlo Cacciabue

Modelling and Identification in Robotics
Krzysztof Kozlowski

Spacecraft Navigation and Guidance
Maxwell Noton

Robust Estimation and Failure Detection
Rami Mangoubi

Adaptive Internal Model Control
Aniruddha Datta

Price-Based Commitment Decisions in the Electricity Market
Eric Allen and Marija Ilic

Compressor Surge and Rotating Stall
Jan Tommy Gravdahl and Olav Egeland

Radiotherapy Treatment Planning
Oliver Haas

Feedback Control Theory For Dynamic Traffic Assignment
Pushkin Kachroo and Kaan Özbay

Control Instrumentation for Wastewater Treatment Plants
Reza Katebi, Michael A. Johnson and Jacqueline Wilkie

Autotuning of PID Controllers
Cheng-Ching Yu

Robust Aeroservoelastic Stability Analysis
Rick Lind and Marty Brenner

Biao Huang and Sirish L. Shah

Performance Assessment of Control Loops

Theory and Applications

With 102 Figures

Springer

Biao Huang, PhD
Sirish L. Shah, PhD
Department of Chemical and Materials Engineering, University of Alberta,
Edmonton, Alberta, Canada, T6G 2G6

ISBN 1-85233-639-0 Springer-Verlag London Berlin Heidelberg

British Library Cataloguing in Publication Data
Huang, Biao, 1962-
  Performance assessment of control loops : theory and
  Applications. - (Advances in industrial control)
  1. Automatic control
  I. Title   II. Shah, S. L. (Sirish L.), 1949-
  629.8
ISBN 1852336390

Library of Congress Cataloging-in-Publication Data
Huang, Biao
   Performance assessment of control loops : theory and applications
 / Biao Huang and Sirish L. Shah.
       p. cm.
    Includes biblipgraphical references.
    ISBN 1-85233-639-0
    1. Automatic control. 2. Control theory. 3. Algorithms.
  I. Shah, S. L. (Sirish L.), 1949- . II. Title.
  TJ213.H83   1999   99-31704
  629.8—dc21   CIP

Apart from any fair dealing for the purposes of research or private study, or criticism or review, as permitted under the Copyright, Designs and Patents Act 1988, this publication may only be reproduced, stored or transmitted, in any form or by any means, with the prior permission in writing of the publishers, or in the case of reprographic reproduction in accordance with the terms of licences issued by the Copyright Licensing Agency. Enquiries concerning reproduction outside those terms should be sent to the publishers.

© Springer-Verlag London Limited 1999
Printed in Great Britain

MATLAB® and SIMULINK® are the registered trademarks of The MathWorks, Inc., http://www.mathworks.com

A significant portion of the material in this book has been reprinted from various journals and appears with permission from Elsevier Science and the Canadian Journal of Chemical Engineering. For the full citations of the papers involved please refer to page 7.

The use of registered names, trademarks, etc. in this publication does not imply, even in the absence of a specific statement, that such names are exempt from the relevant laws and regulations and therefore free for general use.

The publisher makes no representation, express or implied, with regard to the accuracy of the information contained in this book and cannot accept any legal responsibility or liability for any errors or omissions that may be made.

Typesetting: Camera ready by authors
Printed and bound by the Athenæum Press Ltd., Gateshead, Tyne & Wear
69/3830-543210 Printed on acid-free paper SPIN 10728715

Advances in Industrial Control

Series Editors

Professor Michael J. Grimble, Professor of Industrial Systems and Director
Professor Michael A. Johnson, Professor of Control Systems and Deputy Director

Industrial Control Centre
Department of Electronic and Electrical Engineering
University of Strathclyde
Graham Hills Building
50 George Street
Glasgow G1 1QE
United Kingdom

Series Advisory Board

Professor Dr-Ing J. Ackermann
DLR Institut für Robotik und Systemdynamik
Postfach 1116
D82230 Weßling
Germany

Professor I.D. Landau
Laboratoire d'Automatique de Grenoble
ENSIEG, BP 46
38402 Saint Martin d'Heres
France

Dr D.C. McFarlane
Department of Engineering
University of Cambridge
Cambridge CB2 1QJ
United Kingdom

Professor B. Wittenmark
Department of Automatic Control
Lund Institute of Technology
PO Box 118
S-221 00 Lund
Sweden

Professor D.W. Clarke
Department of Engineering Science
University of Oxford
Parks Road
Oxford OX1 3PJ
United Kingdom

Professor Dr -Ing M. Thoma
Institut für Regelungstechnik
Universität Hannover
Appelstr. 11
30167 Hannover
Germany

Professor H. Kimura
Department of Mathematical Engineering and Information Physics
Faculty of Engineering
The University of Tokyo
7-3-1 Hongo
Bunkyo Ku
Tokyo 113
Japan

Professor A.J. Laub
College of Engineering - Dean's Office
University of California
One Shields Avenue
Davis
California 95616-5294
United States of America

Professor J.B. Moore
Department of Systems Engineering
The Australian National University
Research School of Physical Sciences
GPO Box 4
Canberra
ACT 2601
Australia

Dr M.K. Masten
Texas Instruments
2309 Northcrest
Plano
TX 75075
United States of America

Professor Ton Backx
AspenTech Europe B.V.
De Waal 32
NL-5684 PH Best
The Netherlands

To Yali, Linda and my parents (B.H.)

"A journey of a thousand miles begins with a single step"

- Chinese Proverb

To my father and my late mother (SLS)

न हि ज्ञानेन सदृशं पवित्रमिह विध्यते ।

".. Right conduct and knowledge are indeed the most esteemed virtues .."

- 38th verse of Chapter 4 of the Bhagwadgeeta

SERIES EDITORS' FOREWORD

The series *Advances in Industrial Control* aims to report and encourage technology transfer in control engineering. The rapid development of control technology has an impact on all areas of the control discipline. New theory, new controllers, actuators, sensors, new industrial processes, computer methods, new applications, new philosophies..., new challenges. Much of this development work resides in industrial reports, feasibility study papers and the reports of advanced collaborative projects. The series offers an opportunity for researchers to present an extended exposition of such new work in all aspects of industrial control for wider and rapid dissemination.

Benchmarking is a technique first applied by Rank Xerox in the late 1970s for business processes. As a subject in the commercial arena, benchmarking thrives with, for example, a European Benchmarking Forum. It has taken rather longer for benchmarking to make the transfer to the technical domain and even now the subject is making a slow headway. A key research step in this direction was taken by Harris (1989) who used minimum variance control as a benchmark for controller loop assessment. This contribution opened up the area and a significant specialist literature has now developed. Significant support for the methodology was given by Honeywell who have controller assessment routines in their process control applications software; therefore, it is timely to welcome a (first) monograph on controller performance assessment by Biao Huang and Sirish Shah to the Advances in Industrial Control series. Industrial engineers, process control engineers and the academic control community should find this structured presentation of this new class of operational techniques invaluable for self-study or as supplementary taught course material.

M.J. Grimble and M.A. Johnson
Industrial Control Centre
Glasgow, Scotland, U.K.
June, 1999

PREFACE

The design of advanced control algorithms has largely preoccupied the control practitioner's efforts. The rationale has been that systems which are difficult to control need advanced optimal, non-linear, adaptive or like control algorithms for better regulation. Although there is a variety of control design techniques, such as L_1, H_2, H_∞, *etc.*, few techniques exist for objective measures of control loop performance or, conversely, measures of the level of difficulty in controlling a process variable from routine operating industrial process data. The control literature is relatively sparse on studies concerned with such proper or formal measures of control loop performance. The purpose of this monograph is to expose the reader to the most recent techniques for controller performance assessment. Specifically, this text ponders questions such as the following: Is your controller healthy? Is it doing its job well? How can one obtain a non-invasive or a model-free assessment of controller performance?

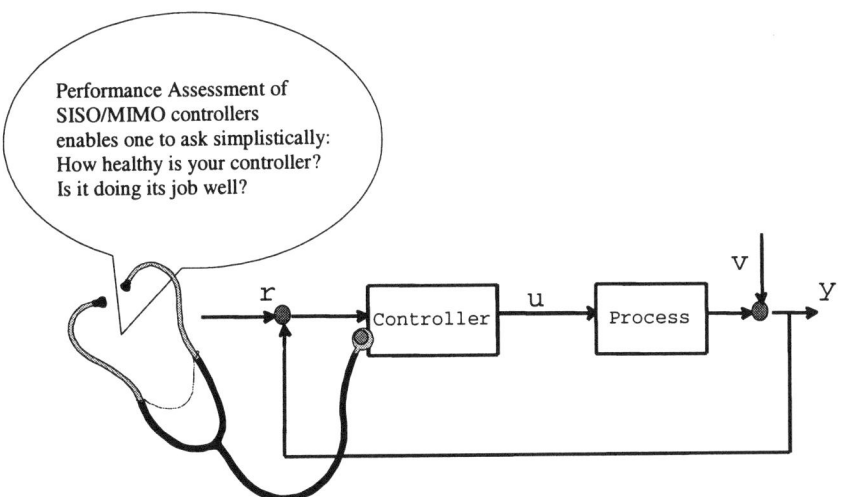

Fig. 0.1. A simplistic depiction of the performance assessment problem.

One of the motivating factors as to why the industrial practitioner and the academic researcher may want to investigate this area is the explosive growth in the amount of process data that is available for analysis. In almost all process environments, ranging from petroleum refining to pulp and paper processing, easy data access through Distributed Control Systems (DCS) is now a rule rather than an exception. In fact, in most process industries data is collected and simply archived. Estimates indicate that most chemical plants require over 100 gigabytes of storage space to archive a year's worth of data. The data-warehousing problem is a manifestation of the exponential increase in information flow as a result of the recent advances in networks and computers. In summary, we live in an information age in which the process industry world is awash with data. Properly archived, screened and sorted data can be a tremendous source of information. The question is how to extract useful information from this data and then put it to good use? The main objective of this research monograph is to suggest how routine closed loop data can be used to obtain non-invasive measures of the performance of the control loops in the plant. The theoretical results are supported by simulation plus experimental (in laboratory) and industrial evaluations.

An equally important motivation for the recent interest in this area is an economic one. A typical industrial process includes thousands of control loops. Instrumentation technicians generally maintain and service these loops, but rather infrequently. Routine maintenance of such loops at optimal settings can save a typical chemical complex hundreds of thousands of dollars a year. The development of quality measures of performance for such control loops in industrial settings is therefore an important area of industrial interest. Controller monitoring falls in the realm of enterprise asset management. The thinking is that controllers, whether basic (*i.e.*, PID type) or advanced, should be treated like other capital assets and they should therefore be monitored on a routine basis.

The following synopsis outlines the major ingredients of this book. Performance assessment of univariate control loops is first carried out by comparing the actual output variance with the minimum variance. The latter term is estimated by simple time series analysis of routine closed-loop operating data. The only prerequisite information required is knowledge of the process delay. When the process has a unit delay, then a whiteness test can be applied to the output as is typically done via the Shewhart chart. The analysis of the same problem in the frequency domain has not received much attention in the literature. Preliminary results in this area are presented here as they complement the time-domain analysis very well and provide an added perspective on the controller performance assessment problem.

The univariate performance assessment concepts are then extended to the multivariate case. A key to performance assessment of multivariate processes, using minimum variance control as a benchmark, is to estimate the benchmark performance from routine operating data with *a priori* knowledge of time-delay or the interactor-matrix. An algorithm for the estimation of the interactor matrix from closed-loop data is also developed. The expression for the feedback controller-invariant (minimum variance) term is then derived by using the unitary, weighted unitary and generalized unitary interactor matrices. It is shown that this term can be estimated from routine operating data. The same idea is extended to the performance assessment of systems with non-invertible zeros and to the performance assessment of multivariate feedback plus feedforward controllers. Although these methods are originally developed for stochastic systems, it is shown that the same methods can also be applied to deterministic systems by appropriate reformulation of the initial problem. Thus, a unified approach for control loop performance assessment is proposed. Efficient algorithms for performance assessment are developed and evaluated by simulations as well as by applications on real industrial processes.

Minimum variance characterizes the most fundamental performance limitation of a system, owing to the existence of time-delays/infinite-zeros. Practically, there are many limitations on the achievable control loop performance. For example, a feedback controller that indicates poor performance relative to minimum variance control is not necessarily a poor controller. Further analysis of other performance limitations with more realistic benchmarks is usually required. Performance assessment in a more practical context, such as a user-defined benchmark or control action constraints, is therefore proposed and evaluated by applications in this book. On the other hand, rigorous performance assessment generally requires complete knowledge of a plant model. An identification effort is usually required. As a complement to existing identification methods, a two-step closed-loop identification method is proposed and tested by simulated and experimental data from a computer-interfaced pilot-scale process.

Most of the results presented here have appeared in the archival literature over the last several years and this book attempts to consolidate most of these results in one place. In this respect, the book is likely to be of use for the new researcher as a monograph and as a place to look for basic information presented in a tutorial format because space is no longer a restriction as it is in technical papers in archived journals. However, there are also results here that will appeal to the industrial practitioner. Portions of the book will also be of use in a graduate level course in process control, signal processing and system identification. Students in a senior-year elective course in advanced process control may also find one or more chapters of this book useful. For the interested reader/researcher the "flow-diagram" shown in Figure 0.2 is suggested for delving into this book.

where

Topics	Chapter
Introduction: motivation and basic preliminaries	1
Univariate Performance Assessment Feedback performance assessment Univariate-performance assessment using user-defined benchmarks Industrial univariate case studies Spectral techniques in performance assessment	2 12 16 C
Multivariate Performance Assessment Preliminaries Interactor matrices and their estimation Multivariate feedback performance assessment Multivariate feedback plus feed-forward performance assessment Performance assessment of non-minimum phase systems Performance assessment for stochastic and deterministic disturbances Multivariate performance assessment using user-defined benchmarks Industrial case studies: multivariate case	3 4-5 6-8 9 10 11 12 16
New Topics Performance assessment: LQG benchmark Performance monitoring via model validation Closed loop identification	13 14 15

Fig. 0.2. Suggested reading flow-diagram of the book.

A { Materials of interest to
Senior undergraduate students
Graduate students
Process control instructors
Industrial practitioners

B { Materials of interest to
Graduate students
Graduate level course instructors
Researchers

The material in this book has been the outcome of several years of research efforts by the authors and many other graduate students at the University of Alberta. Several chapters are also authors' most recent work. For example, Chapter 16 is the most recent contributions owing to the authors' experiences in the industrial application of the performance assessment theory. Chapter 14 on model validation and loop monitoring is the outcome of the most recent contribution owing to the collaborative research work between the first author and Edgar C. Tamayo of Syncrude Canada Ltd. Vishnubhotla and Badmus have contributed to the industrial case studies in Chapter 16. Appendix C on spectral techniques in performance assessment is the contribution owing to Vishnubhotla. It is a pleasure to be able to thank the many people who have contributed so generously in the conception and creation of the material in this text. The Computer Process Control Group at the Department of Chemical and Materials Engineering, University of Alberta, "lives" in a highly stimulating environment where creativity and innovation are encouraged. The camaraderie and the cooperative spirit within the group have allowed us to debate and discuss our ideas freely. The broad range of talent within this group has also allowed cross-fertilization and nurturing of many different ideas that have invariably influenced the direction of this book. We are naturally indebted to all members, past and present, of the group. We would particularly like to thank Grant Fisher, Reg Wood and Fraser Forbes for their valuable suggestions and help with our research work, along with past and present graduate students who have shaped our research in one way or another. In particular, we want to thank Lanre Badmus, Pranob Banerjee, Hiroyuki Fujii, Ravindra Gudi, S. (Laksh) Lakshminarayanan, Randy Miller, Rohit Patwardhan, Anand Vishnubhotla, Amy Yiu, Ramesh Kadali, Bhushan Gopaluni, Xin Huang and Ashish Malhotra. We are also indebted to Ezra Kwok, Hiroyuki Fujii, Edgar C. Tamayo and Jim Zurcher for cooperating so enthusiastically and providing us with an opportunity for the industrial application and evaluation of our research results. We also like to thank Oliver Jackson and Ruth Jordan for their editorial assistance in the final editing of this book.

LIST OF FIGURES

0.1	A simplistic depiction of the performance assessment problem.	xi
0.2	Suggested reading flow-diagram of the book.	xiv
2.1	Schematic diagram of SISO process under feedback control.	11
2.2	Schematic diagram of the FCOR algorithm	13
2.3	Schematic diagram of white noise or innovation sequence estimation	14
2.4	Comparison of the ARMA, R^2 and FCOR approaches for control loop performance measures	16
2.5	Schematic diagram of the industrial cascade reactor control loop.	16
2.6	Estimation of minimum variance and performance measure for the inner loop	17
2.7	Estimation of minimum variance and performance measure for the outer loop	18
5.1	A simplified process control loop diagram	44
5.2	Schematic of the two-interacting-tank pilot-scale process.	50
5.3	Open-loop (input and output) test data where $u = 0$ corresponds to 50% opening of the valve, and the units of h_1 and h_2 are voltage. The time scale is in terms of sampling intervals.	51
5.4	Closed-loop (dither and output) test data where the units of h_1 and h_2 are voltage. The time scale is in terms of sampling intervals.	52
5.5	The industrial process flowsheet	52
5.6	Setpoints for the closed-loop tests. The time scale is in terms of sampling intervals	53
5.7	Predicted and actual outputs; all data have been zero-mean centered. The time scale is in terms of sampling intervals	54
6.1	Simple interactor matrix MIMO performance assessment. Each asterisk represents an estimation based on 1000 data points using the FCOR algorithm.	63
7.1	Schematic diagram of the headbox	72
7.2	Schematic diagram of the control system	74
7.3	Process data trajectory	76

7.4 Performance assessment from the single-input and single-output approach. 77
7.5 Performance assessment from the multivariate approach. 79

8.1 Performance assessment of a square MIMO process (with a general interactor matrix) under multiloop minimum variance control 92
8.2 Performance assessment of a square MIMO process (with a general interactor matrix and output weighting) under multiloop minimum variance control 93
8.3 Performance assessment of a non-square MIMO process (with a general interactor matrix) under multivariable control 94
8.4 Schematic diagram of the industrial absorption process 95
8.5 Absorption process data 96
8.6 Multivariable performance assessment of absorption process 96

11.1 Block diagram of closed-loop process 122
11.2 Probability distribution of the shock 123
11.3 Signal generated by passing the shock through filters 124
11.4 Block diagram representation of the simulated process 129
11.5 Process response and identification results (Dahlin controller) 130
11.6 Predicted output response to a step disturbance (Dahlin controller) 131

12.1 Control loop configuration under the IMC framework 141
12.2 Closed-loop impulse response coefficients for a simple integral controller. 141
12.3 Closed-loop impulse response coefficients for an integral plus filter controller. 142
12.4 Closed-loop impulse response coefficients for a detuned integral plus filter controller. 142
12.5 Block diagram of the closed-loop system. 149

13.1 An example of the tradeoff curve; this tradeoff curve separates achievable and non-achievable performance regions............ 154
13.2 Schematic diagram of the pilot-scale process. 159
13.3 Time domain validation of the open-loop model. The time scale is in terms of sampling intervals. 160
13.4 Closed-loop test for controller #1. The time scale is in terms of sampling intervals. 163
13.5 Closed-loop test for controller #2. The time scale is in terms of sampling intervals. 163
13.6 Closed-loop test for controller #3. The time scale is in terms of sampling intervals. 164
13.7 Closed-loop test for controller #4. The time scale is in terms of sampling intervals. 164

13.8 Tradeoff curve estimated from different sets of data. Controller #3 (DMC#3) is not shown in the graph for clarity of the graph. . 165
13.9 Time domain validation for controller #1. The time scale is in terms of sampling intervals. 165
13.10 Performance assessment of the four controllers. 166
13.11 Schematic diagram of the industrial cascade reactor control loop . 166
13.12 Performance assessment of the industrial cascade control loop (outer loop) 168

14.1 Schematic of time-variant process 170

15.1 Process model block diagram 182
15.2 Feedback control loop block diagram 184
15.3 Equivalent transformation of block diagrams. 194
15.4 Residual test for the model identified by using direct identification method (first-order plant and third-order noise). 198
15.5 Residual test for the model identified by using the y-filtering method (first-order plant and third-order noise). 199
15.6 Comparison between direct identification and the y-filtering methods. 199
15.7 Residual test for the model identified by using the y-filtering method (first-order plant and fifth-order noise). 200
15.8 Residual test for the model identified by using the y-filtering method (second-order plant and second-order noise). 201
15.9 Variance of the estimate calculated from Monte-Carlo simulation (second-order plant and second-order noise). 202
15.10 Predicted 1σ bound of the Nyquist plot (second-order plant and second-order noise). 203
15.11 Comparison between y-filtering and w-filtering approaches. 203
15.12 Cross-correlation test for w-filtering. 204
15.13 Cross-correlation test for y-filtering. 204
15.14 Estimate of the sensitivity function 204
15.15 The upper plot is the sensitivity function. The lower plot is the averaged Bode magnitude graph of $\hat{T}$ over 50 runs. 205
15.16 The averaged Nyquist plot of the estimate over 50 runs 205
15.17 Schematic of the computer-interfaced pilot-scale process. 207
15.18 Block diagram for implementation of IMC control using the real-time SIMULINK Workshop. 207
15.19 Excitation signal and response. The physical units are voltage in the plot where $-2V$ to $+2V$ correspond to 0% to 100%. The time scale is in terms of sampling intervals. 208
15.20 Predicted and actual data from another open-loop test. The time scale is the sampling intervals. 208

15.21 Excitation signal and response under the closed- loop condition. All physical units are voltage in the plot where $-2V$ to $+2V$ correspond to 0% to 100%. The time scale is in terms of sampling intervals. 209
15.22 Comparison of the identified process models using different methods when a second-order model is used. 210
15.23 Comparison of the identified process models using different methods when a first-order model is used. 210
15.24 Effect of the shaping filter for the first-order model. 211

16.1 Impulse response subject to impulse disturbance................ 214
16.2 Typical process data and corresponding performance indices 215
16.3 Autocorrelation function 216
16.4 Impulse responses estimated from routine operating data 217
16.5 Frequency response estimated from routine operating data....... 218
16.6 Autocorrelation of residuals................................... 219
16.7 Segmentation of performance indices 219
16.8 Reference benchmarking based on impulse responses 221
16.9 User specified benchmark based on impulse responses 221
16.10 Multivariate performance assessment of a simulated distillation column control system 222
16.11 Autocorrelation function of multivariate process 223
16.12 Frequency response of multivariate process 224
16.13 Normalized multivariate impulse response..................... 226
16.14 Auto-correlation function of the composition controller 227
16.15 Performance index for a range of delays (1 - 5 minutes) 227
16.16 Performance index for a moving window of size 180 samples 228
16.17 Closed loop impulse response coefficients of the composition controller response (PI=0.45, d=3 min) 229
16.18 Closed loop frequency response plots of the composition controller response (PI=0.45, d=3 min.) 229
16.19 Comparisons of the closed impulse response plots of the composition controller response before and after tuning 230
16.20 Comparisons of the closed loop frequency response plots of the composition controller response before and after tuning 231
16.21 Pressure and level data with sampling interval 5 seconds 232
16.22 Individual output performance indices 233
16.23 Normalized multivariate impulse response..................... 233
16.24 Auto and cross correlation of process output 234
16.25 Frequency domain analysis of process output 234

C.1 Comparison of performance through spectrum analysis. 245

LIST OF TABLES

9.1 The procedure for calculation of the minimum variance feedforward & feedback control benchmark performance of MIMO processes 104

10.1 The procedure for calculation of the benchmark performance of MIMO processes with non-minimum phase zeros 116

13.1 The procedure for calculation of the LQG tradeoff curve 157
13.2 Controller tuning parameters 160

14.1 Summary of Monte-Carlo detection simulation for 10% parameter change using ARX modeling 177
14.2 Summary of Monte-Carlo detection simulation for 20% parameter change using ARX modeling 177
14.3 Summary of Monte-Carlo detection simulation for 10% parameter change using OE modeling 177
14.4 Summary of Monte-Carlo detection simulation for 20% parameter change using OE modeling 178

15.1 Expressions of the asymptotic variance and bias errors 185
15.2 Item to item correspondence between two equations 187
15.3 The procedure for two-step identification 191
15.4 The procedure for two-step identification plus shaping 191

NOMENCLATURE

a_t, a_i	white noise sequences (as driving force of disturbances)
$\hat{a}_t$	estimated white noise sequence
b_t	driving force of measurable disturbances
c	constant
D	interactor matrix
D_w	weighted unitary interactor matrix
D_G	generalized unitary interactor matrix
$\underline{D}$	algebraic form of the interactor matrix
D_f	interactor matrix that contains finite unstable zeros
D_{inf}	interactor matrix that contains infinite zeros
DMC	dynamic matrix control
d	order of interactor matrix (MIMO), time-delay (SISO)
d_i	time-delay in the ith output
$diag(.)$	diagonal matrix operator
e_t	process output under minimum variance control
$\tilde{e}_t$	interactor-filtered e_t
$E(.)$	expectation (mean) operator
$F, \tilde{F}$	matrix polynomials
$F_i, \tilde{F}_i$	constant matrices
FCOR	Filtering and Correlation analysis
G_i	Markov parameter matrix
$\underline{G}$	block-Toeplitz matrix
G_f, G_F	filter transfer function (matrix)
G_R	desired closed-loop transfer function (matrix)
GPC	generalized predictive control
I	identity matrix
IMC	internal model control
J	index of linear quadratic objective function
K	full rank constant matrix
K_t	rush : drag ratio
K^{sp}	setpoint of rush : drag ratio
$L_i, \tilde{L}_i$	constant matrices
LQ	linear quadratic
m	column dimension of transfer function matrix T

M	number of data points
M_F	$= q^d F$, polynomial matrix
MIMO	multi-input and multi-output
n	row dimension of transfer function matrix T
N	disturbance transfer function (matrix for MIMO system)
$\tilde{N}$	interactor-filtered disturbance transfer function matrix
$\hat{N}$	model used to identify disturbance dynamics
N_{ij}, N'_{ij}	disturbance transfer functions
N_a	transfer function matrix of unmeasurable disturbances
N_b	transfer function matrix of measurable disturbances
OEM	output error method
p_t	air pad pressure (deviation variable) (Pa)
p_t^{sp}	setpoint of air pad pressure (Pa)
P_t	air pad pressure (original variable) (Pa)
PEM	prediction error method
q^{-1}	backshift operator
Q	feedback controller transfer function
Q^*	IMC controller transfer function (matrix)
r	number of infinite zeros
$R, \tilde{R}$	rational transfer function matrices
R_{mp}	stable polynomial transfer function (matrix)
R_{nmp}	unstable polynomial transfer function (matrix)
S	sensitivity function (matrix)
SISO	single-input and single-output
SSE	sum of square error
SVD	singular value decomposition
t_d	process time-delay (s)
t_c	control interval (s)
t_s	sampling interval (s)
$tr(.)$	trace of matrix
T	process transfer function (matrix for MIMO system)
$\tilde{T}$	delay-free transfer function matrix
$\hat{T}$	model used to identify plant dynamics
T_{ij}, T'_{ij}	transfer functions
u_t, U_t	manipulated variables
u_1	manipulated variable of air flow control valve
u_2	manipulated variable of fan pump speed
V_t	wire speed (m/s)
$w_{t-d}, \tilde{w}_{t-d}$	portion of output due to non-optimal FB
w	dither signal
W	weighting matrix
x	process input
y_i	individual output
$\tilde{y}_i$	interactor-filtered individual output

Y_t	process output (vector)
$\tilde{Y}_t$	interactor-filtered process output (vector)
Y_t^{sp}	process setpoint (vector)
$Y_t\|_{mv}$	output under MV control
$\tilde{Y}_t\|_{mv}$	filtered output under MV control
Z	block matrix of correlation coefficient matrices

Greek

$\eta(d)$, $\eta'(d)$	performance indices (PI)
η_{y_i}	the ith individual performance index
η_{min}	PI with MV or opt. H_2 control as benchmark
η_{admv}	PI with admissible MV or opt. H_2 control as benchmark
η_{user}	PI with user-specified control as benchmark
ω	frequency (rad/s)
σ_{mv}^2	minimum variance
$\tilde{\sigma}_{mv}^2$	achievable minimum variance
Σ_{mv}, $\Sigma_{\tilde{mv}}$	minimum variance matrix
Σ_a	variance matrix of a_t
$\tilde{\Sigma}_a$	diagonal matrix of Σ_a
σ_y^2	output variance
Σ_Y	output variance matrix
$\Sigma_{\tilde{Y}}$	interactor-filtered output variance matrix
$\tilde{\Sigma}_Y$	diagonal matrix of Σ_Y
$\tilde{\Sigma}_{\tilde{Y}}$	diagonal matrix of $\Sigma_{\tilde{Y}}$
$\Sigma_{\tilde{Y}a}(i)$	covariance matrix between $\tilde{Y}_t$ and a_{t-i}
ρ_a	autocorrelation matrix of a_t
$\rho_{\tilde{Y}a}(i)$	cross correlation matrix between $\tilde{Y}_t$ and a_{t-i}
$\bar{\rho}_{\tilde{Y}a}(i)$	scaled cross correlation matrix
ϵ_t	tracking error
ν_t	white noise sequence
θ	rational transfer function
ξ_t	measured disturbances

TABLE OF CONTENTS

1. Introduction 1
   1.1 An overview of control loop performance assessment 1
   1.2 Objectives of this book 5
   1.3 Organization of the book 6

2. Feedback Controller Performance Assessment of Univariate Processes 9
   2.1 Introduction 9
   2.2 Feedback controller-invariance of minimum variance term ... 10
   2.3 The FCOR algorithm for SISO processes 12
       2.3.1 The FCOR algorithm 12
       2.3.2 Filtering or whitening 13
   2.4 Evaluation via simulation and industrial application 15
   2.5 Summary 18

3. Multivariate Processes: Preliminaries 19
   3.1 Introduction 19
   3.2 Preliminaries of MIMO processes 19
   3.3 Summary 22

4. Unitary Interactor Matrices and Minimum Variance Control 23
   4.1 Introduction 23
   4.2 Unitary interactor matrices 24
   4.3 Unitary interactor matrices and the minimum variance control law 26
   4.4 Weighted unitary interactor matrices and singular LQ control 30
   4.5 Numerical Example 34
   4.6 Summary 36

5. Estimation of the Unitary Interactor Matrices 37
   5.1 Introduction 37
   5.2 Determination of the order of interactor matrices 39
   5.3 Factorization of unitary interactor matrices 41

    5.4   Estimation of the interactor matrix under closed-loop conditions 43
    5.5   Numerical rank 46
    5.6   Simulation and experimental evaluation on a pilot scale process 47
    5.7   Industrial application 51
    5.8   Summary.. 55

6. **Feedback Controller Performance Assessment: Simple Interactor** 57
    6.1   Introduction 57
    6.2   Feedback controller-invariance of minimum variance term.... 57
    6.3   The FCOR algorithm 60
          6.3.1   Multivariable performance index 60
    6.4   Simulation... 62
    6.5   Summary.. 64

7. **Feedback Controller Performance Assessment: Diagonal Interactor** 65
    7.1   Introduction 65
    7.2   Feedback controller-invariance of minimum variance term.... 65
          7.2.1   Feedback controller-invariance of minimum variance term... 65
          7.2.2   Effect of non-minimum phase zeros 68
    7.3   Performance measures 69
          7.3.1   The FCOR algorithm 69
          7.3.2   The effect of sampling intervals 70
    7.4   Application to a headbox control system 71
          7.4.1   Process description 71
          7.4.2   Process control................................... 72
          7.4.3   Problem description 73
    7.5   Performance assessment of the headbox control system 74
          7.5.1   Single loop performance assessment................. 74
          7.5.2   Multivariate performance assessment................ 77
    7.6   Summary.. 80

8. **Feedback Controller Performance Assessment: General Interactor** 81
    8.1   Introduction 81
    8.2   Feedback controller-invariance of minimum variance term.... 82
          8.2.1   Review of the unitary interactor matrix 82
          8.2.2   Feedback controller-invariance of minimum variance term and its separation from routine operating data .. 82
    8.3   The FCOR algorithm for general interactor matrices 88
          8.3.1   Multivariable performance measures 88
    8.4   Evaluation of the FCOR algorithm on a simulation and an industrial application 90

	8.4.1 Simulated example 90

        8.4.1   Simulated example 90
        8.4.2   Industrial application 94
   8.5 Summary... 97

9. **Feedforward & Feedback Controller Performance Assessment** 99
   9.1 Introduction 99
   9.2 Feedforward & feedback controller performance assessment of MIMO processes 99
        9.2.1   Minimum variance FF&FB control benchmark performance 99
        9.2.2   Feedback controller performance assessment of MIMO processes using minimum variance FF & FB control as the benchmark 103
   9.3 Numerical example 104
   9.4 Summary.. 106

10. **Performance Assessment of Nonminimum Phase Systems** . 107
    10.1 Introduction 107
    10.2 Generalized unitary interactor matrices.................. 108
    10.3 Performance assessment of MIMO non-minimum phase processes 111
         10.3.1 Performance assessment with admissible minimum variance control as the benchmark 111
         10.3.2 Alternative proof of admissible minimum variance control 114
    10.4 Numerical example 117
    10.5 Summary.. 120

11. **A Unified Approach to Performance Assessment** 121
    11.1 Introduction 121
    11.2 Setpoint tracking problem 121
    11.3 Deterministic disturbances occurring at random time 123
    11.4 Performance assessment with both stochastic and deterministic disturbances 125
    11.5 Performance assessment with pure deterministic disturbances 126
    11.6 Unified assessment of stochastic and deterministic systems ... 127
    11.7 Simulation.. 128
    11.8 Summary.. 131

12. **Performance Assessment: User-specified Benchmark** 133
    12.1 Introduction 133
    12.2 Preliminaries... 134
    12.3 User-specified performance benchmark: minimum phase systems 134

        12.3.1 SISO case 134
        12.3.2 MIMO case 140
    12.4 User-specified performance benchmark: nonminimum phase
         systems 146
    12.5 Summary.. 151

13. **Performance Assessment: LQG Benchmark** 153
    13.1 Introduction 153
    13.2 Performance assessment with control action taken into account 154
        13.2.1 LQG solution via state space or input-output model .. 154
        13.2.2 LQG solution via GPC.............................. 155
        13.2.3 The tradeoff curve................................. 156
        13.2.4 Performance assessment 157
    13.3 Pilot-scale experimental evaluation 159
    13.4 Case study on an industrial process 162
    13.5 Summary.. 168

14. **Control Loop Performance Monitoring via Model Validation** 169
    14.1 Introduction 169
    14.2 Problem formulation and the local detection approach....... 170
    14.3 Normalized residuals and primary residuals 171
    14.4 Derivation of primary and normalized residuals - an illustra-
         tive example 173
    14.5 Performance monitoring in the presence of varying distur-
         bance dynamics 175
    14.6 Summary.. 178

15. **Closed-loop Identification** 179
    15.1 Introduction 179
    15.2 Accuracy aspects of closed-loop identification 181
    15.3 Two-step closed-loop identification........................ 186
        15.3.1 Estimation of the sensitivity function—step 1 187
        15.3.2 Estimation of the process model—step 2 188
        15.3.3 Other practical considerations...................... 193
    15.4 Extension to MIMO systems 195
    15.5 Simulation.. 195
    15.6 Experimental evaluation on a pilot-scale process 206
    15.7 Summary.. 211

16. **Practical Considerations and Industrial Case Studies** 213
    16.1 Introduction 213
    16.2 Practical considerations in univariate performance assessment 214
    16.3 Univariate performance assessment using alternative perfor-
         mance indicators 216

16.4 Univariate performance assessment using user specified benchmarks 220
16.5 Performance assessment of multivariate control loops 222
    16.5.1 Performance assessment of multivariate control loops using minimum variance control as benchmark 222
    16.5.2 Performance assessment and diagnosis of multivariate control loops using alternative performance indicators . 223
16.6 Industrial case studies of continuous performance assessment . 225
    16.6.1 Composition control loop performance assessment 225
    16.6.2 Capacitance drum control loops performance assessment 230
16.7 Summary 233

Appendices 237

A. The algorithm for the calculation of a unitary interactor matrix 237

B. Examples of the diagonal/general interactor matrices 239

C. Spectral techniques in performance assessment 241
    C.1 Introduction 241
    C.2 Discrete Fourier Transform 241
    C.3 Power Spectrum 242
    C.4 Controller tuning 243
    C.5 Simulation example 244

References 247

Index 253

CHAPTER 1
INTRODUCTION

1.1 An overview of control loop performance assessment

Astrom (1970), Harris (1989), and Stanfelj *et al.* (1993) have reported the use of minimum variance control as a benchmark standard against which to assess control loop performance. DeVries and Wu (1978) have applied the analysis of dispersion and spectral methods to multivariate performance assessment. The most notable work is that by Harris, who, in a 1989 study, showed how simple time series analysis techniques could be used to find a suitable expression for the feedback controller-invariant term from routine operating data of the SISO process and the subsequent use of this as a benchmark to assess control loop performance. This contribution by Harris was significant in the sense that it marked a new direction and framework for the control-loop-performance-monitoring area.

More recently, another related performance assessment statistic defined as the normalized performance index has been proposed by Desborough and Harris (1992). Kozub and Garcia (1993, 1997) have also reported yet another, but similar measure of performance which they define as closed loop potential (or CLP). Lynch and Dumont (1996), Desborough and Harris (1994), Jofriet and Bialkowski (1996), Jofriet *et al.* (1996), Harris *et al.* (1996), and Owen *et al.* (1996) have applied the performance monitoring scheme to pulp/paper processes. Applications of performance assessment techniques in refineries can also be found in Thornhill *et al.* (1996,1998). Tyler and Morari (1995) have extended the same idea to SISO processes with non-minimum phase and/or unstable poles. Eriksson and Isaksson (1994), Rhinehart (1995), Hagglund (1995), Miao and Seborg (1995), and Tyler and Morari (1995) have suggested alternative performance assessment and monitoring schemes for practical consideration. Control loop performance has also been studied from frequency domain analysis(Kendra and Cinar 1997). Huang *et al.* (1995a,1996a) and Harris *et al.* (1995,1996) have extended Harris' performance assessment concepts to performance assessment of MIMO feedback controllers. Kesavan and Lee (1997) have proposed a performance diagnosis tool for multivariate model predictive control systems. An excellent overview of the research in this area can be found in Harris *et al.* (1999) and in Qin (1998). Most recent work on control loop performance assessment has been toward performance monitoring of nonlinear and/or time-variant processes(Huang 1999b), model validation through detection of abrupt changes(Huang 1999a; Huang and Tamayo 1998), and performance monitoring of model predictive control(Huang, Zhao, Tamayo, and Hanafi 1999).

The concept of a delay term is important in minimum variance control. This idea obviously carries over to the MIMO minimum variance control case as well. What is difficult to handle in the MIMO case is the concept of a time-delay matrix (defined elsewhere as the *interactor* matrix (Wolovich and Falb 1976; Goodwin and Sin 1984; Shah, Mohtadi, and Clarke 1987; Tsiligiannis and Svoronos 1988)) as an entity in itself, *i.e.*, one that can be factored out to design a MIMO minimum variance controller if that is the objective. The interactor matrix, as originally proposed by Wolovich and Falb(1976), had a lower triangular form. With this form of the interactor matrix, the minimum variance control law (Goodwin and Sin 1984; Dugard, Goodwin, and Xianya 1984) and minimum ISE control law (Tsiligiannis and Svoronos 1988) are not unique, and, furthermore, are output-order dependent, *i.e.*, under minimum variance control, $Var[y_1(t)]$ is minimized, $Var[y_2(t)]$ is also minimized but subject to the constraint that $Var[y_1(t)]$ is minimized, and so on. Therefore the importance of each output depends on the order it is stacked in the output vector, *i.e.*, the first output variable is the most important for the design of minimum variance control; the last output variable is the least important. Rearrangement of the output variables results in different optimal control law. Nevertheless, the lower triangular interactor matrix has played an important role in classic multivariable control design. Readers are referred to Walgama (1986) and Sripada (1988) for interesting discussions on this issue.

Shah *et al.* (1987) pointed out that selection of the form of an interactor matrix is application-dependent, *i.e.*, it may take an upper triangular form or a full matrix form, and yet in model-based predictive control (MPC) schemes for a specific choice of tuning parameters, this requirement can be avoided. Rogozinski *et al.* (1987) proposed an algorithm for factorization of the nilpotent interactor matrix which has the full-matrix form. Peng and Kinnaert (1992) found the existence of the unitary interactor matrix, which is a special form of the nilpotent interactor matrix. Since the unitary interactor is an all-pass term, factorization of such a unitary interactor matrix does not change the spectral property of the underlying system. This property of the unitary interactor matrix is desirable for minimum variance control or singular LQ control and multivariate control loop performance assessment using minimum variance control as the benchmark (Huang, Shah, and Kwok 1996; Harris, Boudreau, and MacGregor 1996). Here the term "singular LQ control" denotes LQ design without penalty on control action. The minimum variance control law, as developed by Goodwin and Sin (1984) , requires a simple design procedure and is suitable for derivation of the feedback controller-invariant term which is the benchmark of multivariate performance assessment. The downside of this control law is that it is not unique and is input-output order dependent. By introducing the unitary and the weighted unitary interactor matrix into this design procedure, it can be shown that the minimum variance control law is unique and is identical to the singular LQ control law as developed by Harris and MacGregor (1987).

The algorithm for factoring the lower triangular interactor matrix, as suggested by Wolovich and Falb (1976) and Goodwin and Sin (1984), generally requires a complete knowledge of the transfer function matrix. Shah et al. (1987) and Mutoh and Ortega (1993), however, have suggested a solution of the interactor matrix by solving a set of linear, algebraic equations of certain Markov parameter matrices (impulse response coefficient matrices). This latter approach directly connects the Markov parameter matrices to the interactor matrix without going through the transfer function and is numerically convenient and attractive for the estimation of the interactor matrix of a MIMO process. For closed-loop control loop performance assessment, the estimation of the interactor matrix under closed-loop conditions is desired. In this book, an algorithm for the estimation of the unitary interactor matrix is proposed. Using the proposed method, the interactor matrix can be estimated from closed-loop data without the estimation of the open loop transfer function matrix. With complete knowledge of process dynamics, many possible limitations on the achievable performance may be calculated via optimization procedures suggested by Boyd and Barratt (1991) and Dahleh and Diaz-Bobillo (1995).

However, having to know the complete model of a process is not a very attractive approach to process performance monitoring since a typical plant can have hundreds and even thousands of control loops, and identification of all loops is a very demanding requirement. Performance monitoring should be carried out in such a way that the normal production of a process is affected as little as possible. In addition, process dynamics and disturbances may drift from time to time, and the model initially identified may not represent the true dynamics so on-line performance monitoring is necessary.

Different levels of constraints require different levels of process knowledge. Some constraints require less *a priori* knowledge of processes than others. If one can break the constraints into different levels, control loop performance may be assessed from the easiest to the hardest. Only those loops which indicate poor performance at the previous level need be examined at the next level performance assessment. Time-delays pose the most fundamental limitations but typically are relatively easy to obtain or estimate. Therefore, the performance limitation owing to time-delays is assessed at the first level. The second level of performance limitation would be owing to non-invertible zeros. Thus performance assessment of MIMO processes with non-invertible zeros is also discussed in this book.

Minimum variance control is the best possible control in the sense that no other controller can provide a lower output variance. However, its implementation is not recommended in practice owing to its poor robustness and excessive control action. Nevertheless, as a benchmark, it does provide useful information. For example, if a process indicates poor performance relative to minimum variance control, then alternate controller tuning or redesigning of the control algorithm can be considered to improve control loop performance. However, if a process indicates good performance and yet its variance is not within the desired limits, then alternate tuning or redesigning of the control algorithm will not be useful. In this case, alternate control strategies such as feedforward control may be necessary in order to reduce the process variance. Desborough and Harris (1993), Miller and Huang (1997), and Vishnubhotla et al. (1997) have discussed feedforward controller performance assessment of SISO processes. This idea has been extended to the MIMO processes by Huang et al. (1999) and Kadali et al. (1999).

Eriksson and Isaksson (1994) have shown that performance assessment using minimum variance control as a benchmark gives an inadequate measure of the performance if the aim is not stochastic control, but, for example, rejection of deterministic type step disturbances or tracking of setpoints. Tyler and Morari(1995) make a similar claim on this issue. These issues are also considered in this book. It is shown that many practical problems such as those posed by Eriksson and Isaksson and others can be readily solved under the same framework as proposed by Harris (1989) via appropriate formulation of the initial problem. It is also shown that performance assessment of both stochastic and deterministic systems can be unified under the H_2 framework.

Minimum variance characterizes the most fundamental performance limitation of a system owing to the existence of time-delays. Practically, there are many limitations on the achievable control loop performance. Minimum variance control performance requires minimum effort to estimate (routine operating data plus *a priori* knowledge of time-delays), and therefore serves as the most convenient first-level performance assessment benchmark. Only those loops that indicate poor first-level performance then need to be re-evaluated by higher-level performance assessment. A higher-level performance test usually requires more *a priori* knowledge than the knowledge of time-delays. This book also considers other practical benchmarks which are considered for the higher-level performance assessment.

However, all of the aforementioned methods are concerned with performance assessment, which does not, explicitly, take into account the control effort. In general, tighter quality specifications result in smaller variation in the process output but typically require more control effort. One may therefore be more interested in knowing how far away the control performance is from the "best" achievable performance with the same control effort. For example, the problem may be cast as follows: Given $E[u_t^2] \leq \alpha$, what is $\min\{E[y_t^2]\}$? The solution to this problem is discussed by investigating the LQG (Linear Quadratic Gaussian) design method and considering the classic LQG tradeoff curve.

A prerequisite for control loop performance assessment at a higher level is generally a model of the process. Ideally, this model should be estimated under closed-loop conditions so that it does not upset normal process operation. A new, two-step closed-loop identification algorithm is developed in this book. The estimated model is shown to have asymptotically identical expressions for the bias and variance terms regardless of how the identification run is conducted, *i.e.*, irrespective of open-loop or closed-loop conditions. The estimated model can then be used subsequently for improving existing controller design, or controller re-design or control loop performance assessment, or general analysis.

1.2 Objectives of this book

The purpose of this book is to present, in a unified manner, the extension of the univariate control loop performance results to the multivariate case. An equally important objective is to also include pertinent and current theoretical results in this area. These include

1. Extension of the unitary interactor matrix to the weighted unitary interactor matrix and the generalized unitary interactor matrix.
2. Proof of equivalence between the minimum variance control law (Goodwin and Sin 1984) and the singular LQ control law (a special solution in Harris and MacGregor(1987)) if a weighted unitary interactor matrix is used.
3. Factorization and estimation of the interactor matrix under both open and closed-loop conditions, which is a necessary prerequisite step for control loop performance assessment.
4. Proof of the *feedback controller invariance* of the output minimum variance performance for MIMO systems by using the unitary, weighted unitary or generalized unitary interactor matrices. This is the key to control loop performance assessment.
5. Development of an efficient algorithm for control loop performance assessment involving filtering and correlation analysis (the FCOR algorithm), which simplifies the calculations and allows the new technique to be easily applied to industrial processes.
6. Development of a performance assessment algorithm for MIMO processes with nonminimum phase zeros.
7. Development of a performance assessment algorithm for feedforward plus feedback control.
8. Proposal of a unified approach for control loop performance assessment under both stochastic and deterministic frameworks, and under regulatory and setpoint tracking frameworks.
9. Extension of performance assessment methodology to cover practical situations such as performance assessment with user-defined benchmarks.
10. Proposal of an LQG benchmark which can take control action constraints into account for performance assessment.

11. Development of an algorithm for *closed-loop identification of SISO/MIMO systems*. This is a spin-off from the work on control loop performance assessment and has strong industrial appeal.

This book is also meant to have a practical appeal for the industrial reader. The methods and algorithms developed in this book have been applied and evaluated at several industrial complexes in the Alberta area as well as internationally. These include:

1. Multivariable control system validation for distillation columns at two Mitsubishi Chemical Corporation locations: 1) Kurosaki Plant and 2) Mizushima Plant, Japan.
2. Multivariable control system validation for a heat exchanger, reactor and distillation column at Agrium Inc's (Sherritt Inc.) Redwater Fertilizer complex in Alberta.
3. Benefit analysis for upgrading the existing headbox control of a paper machine at Weyerhauser Canada's Grande Prairie operations in Alberta.
4. Field test of control loop performance assessment in Syncrude Canada Ltd., Cominco Ltd. and Suncor Energy Inc

1.3 Organization of the book

The book is organized as follows. In Chapter 2, the performance assessment algorithm is first introduced for SISO systems. The key to extend the SISO results to the MIMO system is the understanding of the concept of the time delay matrix or the interactor matrix. This concept is introduced in Chapter 3. In Chapter 4, the role of the unitary interactor matrix in minimum variance or singular LQ control design is discussed. The algorithm for estimation of the interactor matrix is established in Chapter 5. The methods for feedback controller performance assessment of MIMO systems are developed in Chapters 6, 7 and 8. This treatment is in ascending degree of difficulty from the simple interactor, through the diagonal interactor, to the general interactor. When the feedback controller indicates good performance relative to minimum variance control, further improvement of performance may require a different control strategy such as feedforward plus feedback control. The benchmark of feedforward plus feedback control is therefore discussed in Chapter 9. Existence of non-invertible zeros affects the achievable performance of the feedback controller. This issue is addressed in Chapter 10. In Chapter 11, the methodology developed in the previous chapters is extended to performance assessment of deterministic disturbance and/or setpoint tracking. A practical performance assessment for a user-defined benchmark is proposed in Chapter 12. Performance assessment with control action taken into account is a relatively unexplored research area, and one possible solution to this problem is discussed in Chapter 13. Performance assessment with a benchmark other than minimum variance control usually requires an identification effort. A new approach to closed-loop identification is developed in Chapter 15.

A significant portion of the material in this book has appeared in archival journals. Chapter 4 is reprinted from Automatica, Vol 33, B. Huang and S.L. Shah, "The Role of the Unitary Interactor Matrix in the Explicit Solution of the Singular LQ Output Feedback Control Problem", pp. 2071-2075, copyright (1997); Chapter 5 is reprinted from Journal of Process Control, Vol 7, B. Huang, S.L. Shah, and H. Fujii, "The Unitary Interactor Matrix and It's Estimation Using Closed-loop Data", pp. 195-207, copyright (1997); Chapter 8 is reprinted from Automatica, Vol 33, B. Huang, S.L. Shah, and E.K. Kwok "Good, Bad or Optimal? Performance Assessment of Multivariable Processes", pp. 1175-1183, copyright (1997); Chapter 12 is reprinted from Journal of Process Control, Vol 8, B. Huang and S.L. Shah, "Practical Issues in Multivariable Feedback Control Performance Assessment", pp. 421-430, copyright (1998); Chapter 15 is reprinted from Journal of Process Control, Vol 7, B. Huang and S.L. Shah, "Closed-loop Identification: a Two Step Approach", pp. 425-438, copyright (1997); all with permission from Elsevier Science. Chapter 7 is reprinted from Canadian Journal of Chemical Engineering, Vol 75, B. Huang, S.L. Shah, K.E. Kwok, and J. Zurcher, "Performance Assessment of Multivariable Control Loops on a Paper-Machine Headbox", pp. 134-142, Copyright 1997, with permission from Canadian Journal of Chemical Engineering. In order to link the different chapters, there may be some overlap and redundancy of material. This is to ensure completeness and cohesiveness and help the reader understand the material easily.

CHAPTER 2
FEEDBACK CONTROLLER PERFORMANCE ASSESSMENT OF UNIVARIATE PROCESSES

2.1 Introduction

A typical industrial process includes thousands of control loops. Instrumentation technicians generally maintain and service these loops, but rather infrequently. It is important for control engineers to have an efficient tool to monitor and assess control loop performance. Monitoring and assessment of control loop performance should not disturb routine operation of the processes or, at least, should be carried out under closed-loop conditions. As pointed out by Eriksson and Isaksson (1994), *"in the short term, such a tool probably has to be a stand-alone unit with its own software that hooks on to and collects data straight from the input of the process computer; in the long term, such a function will be an integral part of any commercial control system"*.

The control literature on studies concerned with such proper or formal measures of control loop performance has been relatively sparse. Harris (1989) has developed an efficient technique for control loop performance assessment using only routine closed-loop operating data. The control objective is to minimize process variance, and minimum variance control is used as the benchmark standard against which to assess current control loop performance. It has been shown (Harris 1989) that for a system with time delay d, a portion of the output variance is feedback control invariant and can be estimated from routine operating data. This is the minimum variance portion. To separate this invariant term, one needs to model the closed-loop output data y_t by a moving average process such as

$$y_t = \underbrace{f_0 a_t + f_1 a_{t-1} + \cdots + f_{d-1} a_{t-(d-1)}}_{e_t} + f_d a_{t-d} + f_{d+1} a_{t-(d+1)} + \cdots$$

where a_t is a white noise sequence and e_t is the portion of the minimum variance control output independent of feedback control (Harris 1989). This portion of minimum variance can be estimated by time series analysis of routine closed-loop operating data, and can be used subsequently as a benchmark measure of theoretically achievable absolute lower bound of output variance to assess control loop performance. Using minimum variance control as the benchmark does not mean that one has to implement such a controller on the actual process. This benchmark control may or may not be achievable in practice depending on process invertibility and other physical constraints on the processes. However, as a benchmark, it provides useful information such as how good the current controller performance is compared to the minimum variance controller and how much "potential" there is to improve controller performance further. If the controller indicates good performance measure relative to minimum variance control, further tuning or re-designing of the control algorithm is neither necessary nor helpful. In this case, if further reduction of process variation is desired, implementation of feedforward control or re-engineering of the process itself may be necessary. On the other hand, if the controller indicates a poor performance measure, further analysis such as process identification and controller re-design may be necessary since the poor performance measure may be due to constraints such as unstable or poorly damped zeros or control action limits.

As a general introduction to feedback controller performance assessment of multivariate or MIMO processes in this book, control loop performance assessment of univariate or SISO processes is discussed in this chapter. This chapter is organized as follows: In Section 2.2 the feedback control invariant term is re-derived. The FCOR (Filtering and Correlation analysis) algorithm for performance assessment of SISO processes is developed in Section 2.3.1. The proposed algorithm is then evaluated by simulation and actual processes in Section 2.4, followed by concluding remarks in Section 2.5.

2.2 Feedback controller-invariance of minimum variance term

For the sake of brevity and convenience, the backshift operator q^{-1} will be omitted throughout this book unless circumstances necessitate its presence. For example, the transfer function $T(q^{-1})$ will be expressed simply as T. With slight abuse of notation, we will not distinguish between the backshift operator q^{-1} and the Z-transform operator z^{-1}.

Consider a SISO process under regulatory control as shown in Figure 2.1, where d is the time-delay, $\tilde{T}$ is the delay-free plant transfer function, N is the disturbance transfer function, a_t is a white noise sequence with zero mean, and Q is the controller transfer function.

It follows from Figure 2.1 that

$$y_t = \frac{N}{1 + q^{-d}\tilde{T}Q} a_t \qquad (2.1)$$

where using the Diophantine identity:

$$N = \underbrace{f_0 + f_1 q^{-1} + \cdots + f_{d-1} q^{-d+1}}_{F} + R q^{-d}$$

2.2 Feedback controller-invariance of minimum variance term

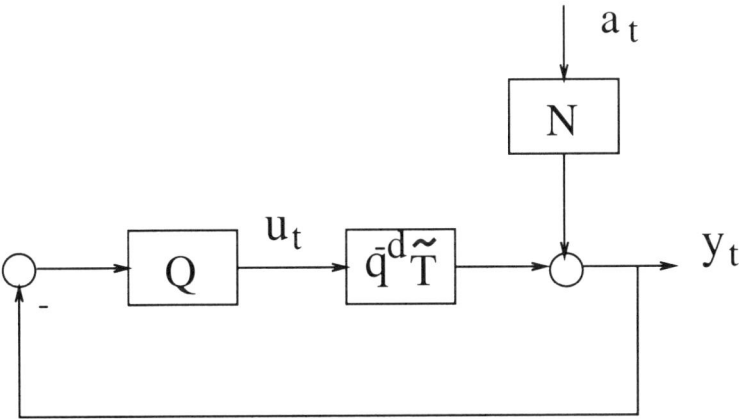

Fig. 2.1. Schematic diagram of SISO process under feedback control.

where f_i (for $i = 1, \cdots, d-1$) are constant coefficients, and R is the remaining rational, proper transfer function, Equation 2.1 can be written as

$$\begin{aligned} y_t &= \frac{F + q^{-d}R}{1 + q^{-d}\tilde{T}Q} a_t \\ &= [F + \frac{R - F\tilde{T}Q}{1 + q^{-d}\tilde{T}Q} q^{-d}] a_t \\ &= F a_t + L a_{t-d} \end{aligned} \qquad (2.2)$$

where $L = \frac{R - F\tilde{T}Q}{1 + q^{-d}\tilde{T}Q}$ is a proper transfer function. Since $F a_t = f_0 a_t + \cdots + f_{d-1} a_{t-d+1}$, the two terms on the right hand side of Equation 2.2 are independent, and as a result,

$$Var(y_t) = Var(F a_t) + Var(L a_{t-d})$$

Therefore

$$Var(y_t) \geq Var(F a_t)$$

The equality holds when $L = 0$, i.e.,

$$R - F\tilde{T}Q = 0$$

This yields the minimum variance control law:

$$Q = \frac{R}{\tilde{T}F}$$

Since F is independent of the controller transfer function Q, the term $F a_t$, which is the process output under minimum variance control, is feedback controller-invariant. Therefore, if a stable process output y_t is modelled by an infinite moving-average model, then its first d terms constitute an estimate of the minimum variance term $F a_t$.

2.3 The FCOR algorithm for SISO processes

2.3.1 The FCOR algorithm

A stable closed-loop process can be written as an infinite-order moving-average (MA) process:

$$y_t = (f_0 + f_1 q^{-1} + f_2 q^{-2} + \cdots + f_{d-1} q^{-(d-1)} + f_d q^{-d} + \cdots) a_t \qquad (2.3)$$

Multiplying Equation 2.3 by $a_t, a_{t-1}, \cdots, a_{t-d+1}$ respectively and then taking the expectation of both sides of the equation yields

$$\begin{aligned}
r_{ya}(0) &= E[y_t a_t] = f_0 \sigma_a^2 \\
r_{ya}(1) &= E[y_t a_{t-1}] = f_1 \sigma_a^2 \\
r_{ya}(2) &= E[y_t a_{t-2}] = f_2 \sigma_a^2 \\
&\vdots \\
r_{ya}(d-1) &= E[y_t a_{t-d+1}] = f_{d-1} \sigma_a^2
\end{aligned} \qquad (2.4)$$

Therefore the minimum variance or the invariant portion of output variance is

$$\begin{aligned}
\sigma_{mv}^2 &= (f_0^2 + f_1^2 + f_2^2 + \cdots + f_{d-1}^2) \sigma_a^2 \\
&= [(\frac{r_{ya}(0)}{\sigma_a^2})^2 + (\frac{r_{ya}(1)}{\sigma_a^2})^2 + (\frac{r_{ya}(2)}{\sigma_a^2})^2 + \cdots + (\frac{r_{ya}(d-1)}{\sigma_a^2})^2] \sigma_a^2 \\
&= [r_{ya}^2(0) + r_{ya}^2(1) + r_{ya}^2(2) + \cdots + r_{ya}^2(d-1)]/\sigma_a^2 \qquad (2.5)
\end{aligned}$$

Let the controller performance index be defined as

$$\eta(d) \triangleq \sigma_{mv}^2 / \sigma_y^2 \qquad (2.6)$$

This has been referred to as the closed-loop potential (CLP) by Kozub and Garcia (1993), and the inequality $0 \leq \eta(d) \leq 1$ is held.

Substituting Equation 2.5 into Equation 2.6 yields

$$\begin{aligned}
\eta(d) &= [r_{ya}^2(0) + r_{ya}^2(1) + r_{ya}^2(2) + \cdots + r_{ya}^2(d-1)]/\sigma_y^2 \sigma_a^2 \qquad (2.7) \\
&= \rho_{ya}^2(0) + \rho_{ya}^2(1) + \rho_{ya}^2(2) + \cdots + \rho_{ya}^2(d-1) \qquad (2.8) \\
&\triangleq ZZ^T \qquad (2.9)
\end{aligned}$$

where Z is the cross-correlation coefficient vector between y_t and a_t for lags 0 to $d-1$ and is denoted as

$$Z \triangleq [\rho_{ya}(0), \rho_{ya}(1), \cdots, \rho_{ya}(d-1)] \qquad (2.10)$$

The corresponding sampled version of the performance index is therefore written as

$$\hat{\eta}(d) = \hat{\rho}_{ya}^2(0) + \hat{\rho}_{ya}^2(1) + \hat{\rho}_{ya}^2(2) + \cdots + \hat{\rho}_{ya}^2(d-1) = \hat{Z}\hat{Z}^T \quad (2.11)$$

where

$$\hat{\rho}_{ya}(k) = \frac{\frac{1}{M}\sum_{t=1}^{M} y_t a_{t-k}}{\sqrt{\frac{1}{M}\sum_{t=1}^{M} y_t^2 \frac{1}{M}\sum_{t=1}^{M} a_t^2}} \quad (2.12)$$

Although a_t is unknown, it can be replaced by the estimated innovations sequence $\hat{a}_t$. The estimate $\hat{a}_t$ is obtained by pre-whitening the process output variable y_t via time series analysis. This pre-whitening procedure will be discussed next. This algorithm is termed the FCOR algorithm for filtering and correlation analysis, and is schematically shown in Figure 2.2.

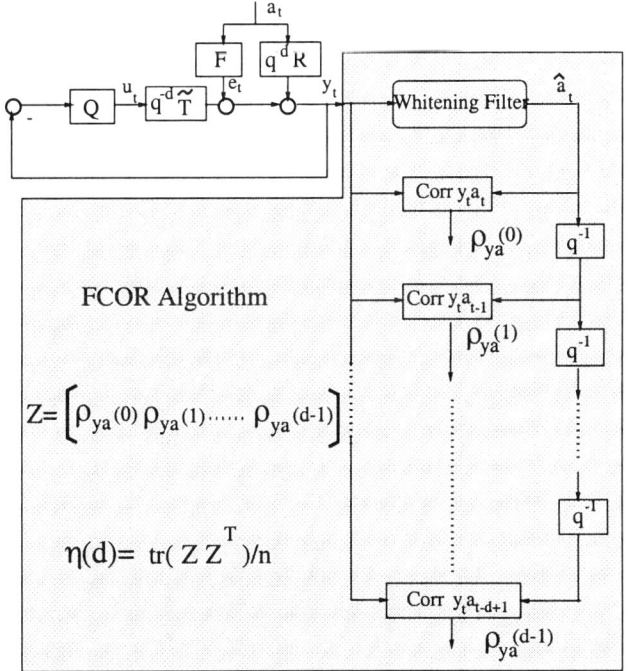

Fig. 2.2. Schematic diagram of the FCOR algorithm

2.3.2 Filtering or whitening

The original source of variation in a regulatory closed-loop process may be traced back to a white noise excitation, a_t, as shown in Figure 2.3. The relationship between y_t and a_t is established by the closed-loop transfer function

$G_{ya} = \frac{N}{1+TQ}$. Thus, the variation of y_t is due to the excitation of a_t through G_{ya}. The estimation of this noise sequence is important for performance assessment. By reversing the process, the white noise sequence can be viewed as an output from a filter whose input is the process output y_t. Many methods have been developed to fit the filter model and obtain estimates of the white noise sequence from output data, and in some literature, the estimation of a_t is known as "whitening" or "prewhitening" (Box and MacGregor 1974; Soderstrom and Stoica 1989; Goodwin and Sin 1984). Such a whitened noise sequence can also be termed the "innovation sequence" (Goodwin and Sin 1984). The process of obtaining such a "whitening" filter is analogous to time-

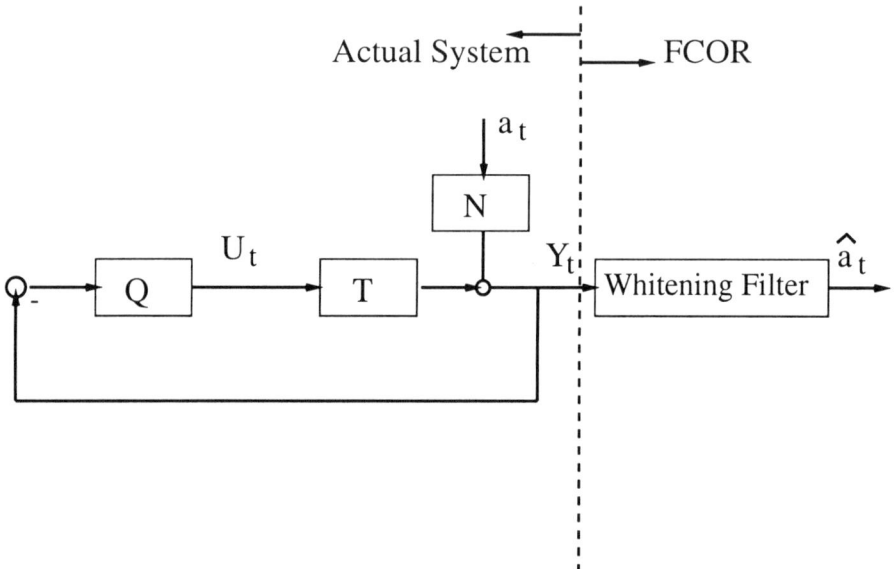

Fig. 2.3. Schematic diagram of white noise or innovation sequence estimation

series modeling, where the final test of the adequacy of the model consists of checking if the residuals are "white". These residuals are the estimated white noise sequence. In contrast to time-series modelling where the estimation of the model is of interest, the residual or the innovation sequence is the main item of interest in this "whitening" process. Depending on data, an AR or ARMA (alternatively a Kalman Filter based innovation model in state space representation) can be used for estimating a_t. The identification of these innovation models (*i.e.*, "whitening" filters) has attracted much interest (Reinsel 1993; Aoki 1987). Many efficient algorithms have been developed such as *arx*, *armax*, etc. in MATLAB (The Math Works, Inc.).

2.4 Evaluation via simulation and industrial application

Example 2.4.1. In order to compare the FCOR algorithm with other available SISO performance assessment algorithms, consider the following SISO process, as used by Desborough and Harris(1992), with time delay $d = 2$:

$$y_t = u_{t-2} + \frac{1 - 0.2q^{-1}}{1 - q^{-1}} a_t \tag{2.13}$$

For a simple integral controller $\Delta u_t = -Ky_t$, it can be shown that the closed-loop response is given by

$$y_t = a_t + 0.8a_{t-1} + \frac{0.8(1 - K/0.8 - Kq^{-1})}{1 - q^{-1} + Kq^{-2}} a_{t-2} \tag{2.14}$$

Note that the first two terms are independent of K and represent the process output under minimum variance control.

The simulation results shown in Figure 2.4 show a comparison of the estimated control performance versus the theoretical performance as a function of K; and a comparison with: 1) the general approach proposed by Harris (1989) (denoted as the ARMA approach here); 2) normalized performance index or R^2 approach (Desborough and Harris 1992); 3) the FCOR algorithm. Desborough and Harris(1992) have used the adjusted multiple coefficient of determination, R^2, as the performance index. This value is converted to the performance index used in this book via the relation, $1 - R^2$.

Example 2.4.2. The proposed performance assessment method was used to assess the performance of an important cascade control loop on a nitric acid (HNO_3) production facility at a world-scale chemical plant in central Alberta, Canada.

The schematic of the process is shown in Figure 2.5. The feed stocks are anhydrous ammonia (NH_3) and air. The ammonia goes through a two-stage heating process before entering the catalytic reaction which contains a (gauze type) platinum-rhodium catalyst. Process air at over 400°F and 150 psig enters the reactor. The ammonia-air mixture reacts on the catalyst at over 1600°F and forms nitrogen dioxide with other by-products (NO_x). In order to maximize the production of NO_2 and minimize the by-products which are harmful to the environment, the gauze temperature is required to be kept as steady as possible even in the presence of disturbances in the ambient temperature air quality, ammonia feed temperature, and ammonia flow rate. The present control configuration is that the gauze temperature controller (outer loop) adjusts the setpoint of the ammonia flow rate (inner loop). In general, the inner loop is tightly tuned and is expected to have a good performance. The time delay of the outer loop from *a priori* analysis is known to be 15s including the delay due to the zero-order-hold device. The sampling period

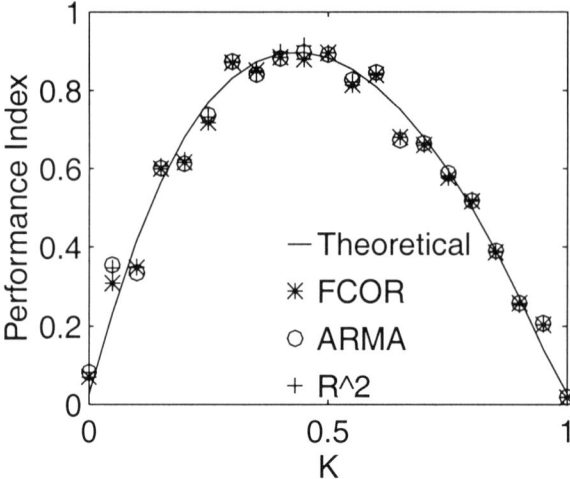

Fig. 2.4. Comparison of the ARMA, R^2 and FCOR approaches for control loop performance measures

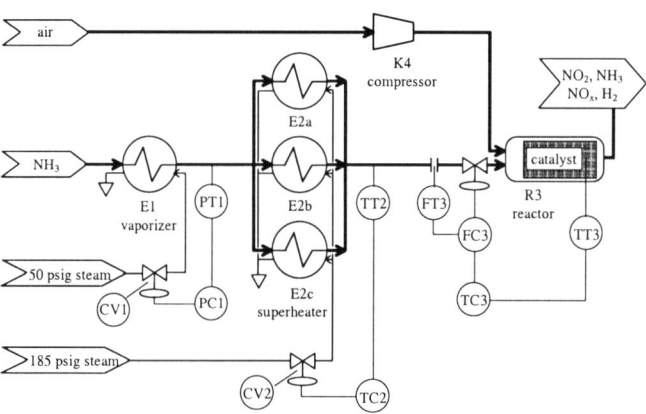

Fig. 2.5. Schematic diagram of the industrial cascade reactor control loop

2.4 Evaluation via simulation and industrial application   17

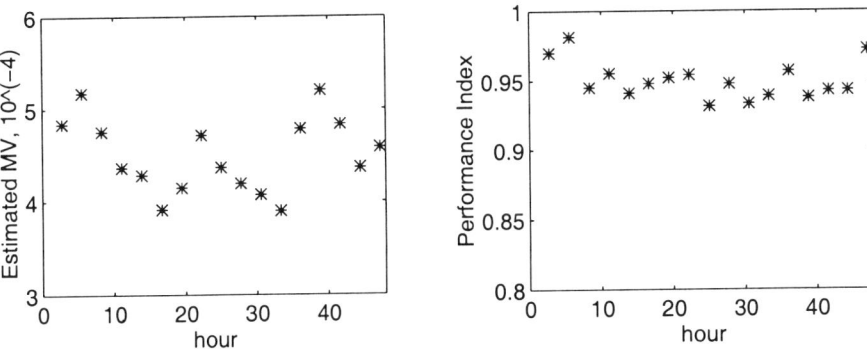

Fig. 2.6. Estimation of minimum variance and performance measure for the inner loop

is 5s; so the time delay is three sampling periods, *i.e.*, $d = 3$. The time delay of the inner loop is considered to be only one sampling interval caused by the zero order hold. A sample size of 35000 points taken over a two-day period is considered. Both loops use PID controllers. The available process data are the gauze temperature, y_1, the outer-loop controller output, which is the setpoint of the inner loop, and the NH_3 flow rate, y_2.

The performance measure of the inner loop by using the FCOR approach is shown in Figure 2.6. On the left part of this figure, each point on the left graph represents the estimated minimum variance or the best achievable output variance based on the calculation of a window of 2000 data points. The right part of Figure 2.6 represents the corresponding performance index estimated using the FCOR approach. The 24-hour periodic trend of the disturbance magnitude is clearly seen from Figure 2.6. Despite this trend, the performance index is close to a constant value of 0.95, which is an indication of excellent performance or loop tuning. Further improvement in this loop by adjusting controller parameters may not be possible.

The estimated minimum variance of the outer loop is shown on the left part of Figure 2.7, and the corresponding performance measure is shown on the right part of this figure. Contrary to what would be expected, the performance index for this loop is not a constant. The trend of the index clearly shows a 24-hour cycling of the loop performance, which is possibly due to the ambient temperature and air quality change over a 24-hour period. The average performance index of the outer loop is approximately 0.15, indicating relatively poor control. Clearly this loop performance may be improved significantly by re-tuning the existing controller and/or providing feedforward control of ambient conditions.

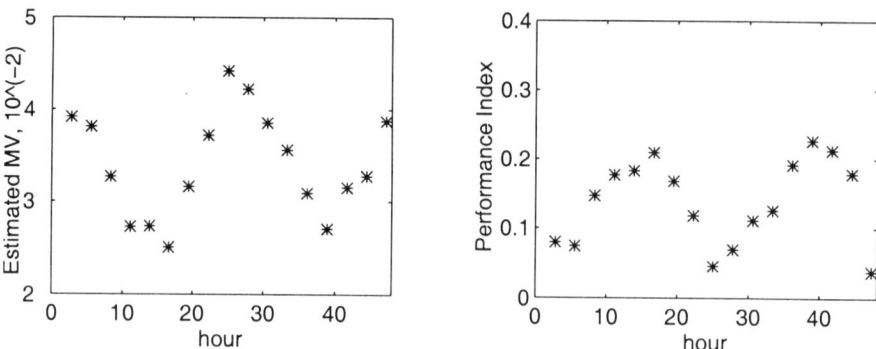

Fig. 2.7. Estimation of minimum variance and performance measure for the outer loop

2.5 Summary

A simple technique for evaluating univariate control loop performance has been proposed. This technique is based on filtering and correlation (FCOR) analysis of routine closed-loop operating data. Use of the proposed method is demonstrated by a simulated and industrial application, and is shown to provide useful insight into control loop performance analysis of univariate processes.

CHAPTER 3
MULTIVARIATE PROCESSES: PRELIMINARIES

3.1 Introduction

Time-delays are the most fundamental limitation on the achievable performance of any feedback controller. Performance assessment of SISO processes, as introduced in Chapter 2, reflects this fundamental performance limitation in the stochastic framework. In the following chapters, we explore performance assessment of multivariable processes. The interactor matrix, a multivariate generalization of the SISO time-delay term, characterizes the most fundamental limitation on the achievable performance of any multivariable feedback controller.

The notion of an interactor matrix (Wolovich and Falb 1976) for a multivariate system can be best understood by relating it to the meaning of the time delay for a univariate process. In the case of a univariate process, the time delay in terms of the sampling time is equal to the number of zero or almost-zero impulse response coefficients, and corresponds to the time that elapses between the moment a change in the input occurs until the moment that this change has an effect on the output or it is the result of the first *nonsingular* or non-zero impulse response coefficient having an effect on the output. From the view point of systems theory, the delay corresponds to the number of infinite zeros of a discrete-time process.

3.2 Preliminaries of MIMO processes

Unless otherwise illustrated, a standard MIMO process model

$$Y_t = TU_t + Na_t \qquad (3.1)$$

is used throughout the book, where T and N are proper (causal), rational transfer function matrices in the backshift operator q^{-1}; Y_t, U_t and a_t are output, input and noise vectors of appropriate dimensions. For stochastic systems, a_t is further assumed to be white noise with zero mean and $Var(a_t) = \Sigma_a$.

To solve the multivariable deadbeat and minimum variance control problems, Wolovich and Falb (1976), Wolovich and Elliott (1983), as well as

3. Multivariate Processes: Preliminaries

Goodwin and Sin (1984) introduced the interactor matrix D, which is the generalization of the SISO time delay for the MIMO case.

Theorem 3.2.1. *For every $n \times m$ proper, rational polynomial transfer function matrix T, there is a unique, non-singular, $n \times n$ lower left triangular polynomial matrix D, such that $|D| = q^r$ and*

$$\lim_{q^{-1} \to 0} DT = \lim_{q^{-1} \to 0} \tilde{T} = K$$

where K is a full rank (full column rank or full row rank) constant matrix, the integer r is defined as the number of infinite zeros of T, and $\tilde{T}$ is the delay-free transfer function (factor) matrix of T which contains only finite zeros. The matrix D is defined as the interactor matrix and can be written as

$$D = D_0 q^d + D_1 q^{d-1} + \cdots + D_{d-1} q$$

where d is denoted as the **order of the interactor matrix** *and is unique for a given transfer function matrix (Shah, Mohtadi, and Clarke 1987; Mutoh and Ortega 1993), and D_i (for $i = 0, \cdots, d-1$) are coefficient matrices.*

The interactor matrix D can be one of the three forms described in the sequel. If D is of the form: $D = q^d I$, then the transfer function matrix T is regarded as having a *simple interactor matrix*. If D is a diagonal matrix, i.e., $D = diag(q^{d_1}, q^{d_2}, \cdots, q^{d_n})$, then T is regarded as having a *diagonal interactor matrix*. Otherwise, T is considered to have a *general interactor matrix* (one realization of which is a triangular interactor matrix). However, the *general interactor matrix* also has forms other than the lower triangular form. It can be a full matrix or an upper triangular matrix (Shah, Mohtadi, and Clarke 1987; Huang, Shah, and Fujii 1996). Rogozinski *et al.* (1987) have introduced an algorithm for the calculation of a *nilpotent interactor matrix*. Peng and Kinnaert (1992) have introduced the *unitary interactor matrix*.

Definition 3.2.1. *Instead of taking the lower triangular form, if an interactor matrix as per Theorem 3.2.1 satisfies*

$$D^T(q^{-1})D(q) = I$$

then this interactor matrix is denoted as the unitary interactor matrix.

The existence of the unitary interactor matrix has been established by Peng and Kinnaert (1992) . To illustrate the point, take a 2×2 transfer function matrix as an example:

$$T = \begin{bmatrix} \frac{q^{-1}}{1+q^{-1}} & \frac{0.5q^{-1}}{1+2q^{-1}} \\ \frac{0.5q^{-1}}{1+3q^{-1}} & \frac{q^{-1}}{1+4q^{-1}} \end{bmatrix}$$

Since

$$\lim_{q^{-1} \to 0} qT = \begin{bmatrix} 1 & 0.5 \\ 0.5 & 1 \end{bmatrix}$$

is a full rank matrix, T has a simple interactor matrix with $D = qI$.

However, if T is changed to

$$T = \begin{bmatrix} \frac{q^{-2}}{1+q^{-1}} & \frac{0.5q^{-3}}{1+2q^{-1}} \\ \frac{0.5q^{-1}}{1+3q^{-1}} & \frac{q^{-1}}{1+4q^{-1}} \end{bmatrix}$$

then

$$\lim_{q^{-1} \to 0} \begin{bmatrix} q^2 & 0 \\ 0 & q \end{bmatrix} T = \begin{bmatrix} 1 & 0 \\ 0.5 & 1 \end{bmatrix}$$

is clearly full rank. Thus the interactor matrix

$$D = \begin{bmatrix} q^2 & 0 \\ 0 & q \end{bmatrix}$$

is a diagonal matrix.

Furthermore, if T is changed to

$$T = \begin{bmatrix} \frac{q^{-1}}{1+q^{-1}} & \frac{q^{-1}}{1+2q^{-1}} \\ \frac{q^{-1}}{1+3q^{-1}} & \frac{q^{-1}}{1+4q^{-1}} \end{bmatrix}$$

then it has a general interactor matrix. Goodwin and Sin (1984) have shown the lower triangular interactor matrix of this last transfer function matrix T as

$$D = \begin{bmatrix} q & 0 \\ -q^3 + 2q^2 & q^3 \end{bmatrix}$$

This can be easily checked by taking the $\lim_{q^{-1} \to 0} DT = K$ and testing that K is full rank. Note that $D^T(q^{-1})D(q) \neq I$. Alternatively, by using the algorithm of Rogozinski et al. (1987), a unitary interactor matrix can be factored out as

$$D = \begin{bmatrix} 0.5q + 0.5q^2 & 0.5q - 0.5q^2 \\ 0.5q^2 - 0.5q^3 & 0.5q^2 + 0.5q^3 \end{bmatrix}$$

This matrix has the property $D^T(q^{-1})D(q) = I$.

A clear explanation of the interactor matrix has been given by Shah et al. (1987). For a SISO transfer function $T = \frac{B}{A} q^{-d}$, the time delay, q^{-d}, introduces d infinite zeros in the transfer function. The transfer function T is not invertible in the sense that the inversion of T, $T^{-1} = \frac{A}{B} q^d$, is not proper. However, if we multiply the transfer function by the interactor matrix (which is a scalar term in the SISO case), $D = q^d$, then the delay-free transfer function $\tilde{T} = DT = \frac{B}{A}$ is invertible in the sense that the inversion is causal or proper. In the MIMO case, the interactor matrix D plays the same role as in the SISO case. Multiplication of the transfer function by the interactor matrix D removes the infinite zeros from the original transfer function matrix T and yields the delay-free transfer function matrix $\tilde{T}$, i.e., $\tilde{T} = DT$.

The introduction of the interactor matrix is important not only because it solves the multivariable minimum variance control problem, but also it provides a basic tool to seek the benchmark performance measure of the multivariable process as we will see in the following sections.

3.3 Summary

Wolovich and Falb's lower-triangular interactor matrix and its extension to the unitary interactor matrix has been introduced in this chapter. Examples have been given to illustrate the concept. The unitary interactor matrix will play a fundamental role in the following chapters.

CHAPTER 4
UNITARY INTERACTOR MATRICES AND MINIMUM VARIANCE CONTROL

4.1 Introduction

There are many limitations to achievable control loop performance, for example, time delays, the existence of poorly damped or non-invertible zeros, constraints on control action, desired robustness characteristics, *etc.* Amongst all these constraints, the time delay is the most fundamental constraint that has attracted much interest in the development of process control. Wolovich and Falb (1976) have shown that the analog of the time-delay term for a SISO system, which is feedback control-invariant, is the interactor matrix for a MIMO system, which is also feedback control-invariant. Subsequently, Wolovich and Elliott (1983) and Goodwin and Sin (1984) extended the concept of the interactor matrix to discrete systems. The interactor matrix characterizes the most fundamental performance limitation of a linear multivariable system.

The interactor matrix, as originally proposed by Wolovich and Falb (1976), had a lower triangular form. With this form of the interactor matrix, the minimum variance control law (Goodwin and Sin 1984; Dugard, Goodwin, and Xianya 1984) and minimum ISE (integral of the squared error) control law (Tsiligiannis and Svoronos 1988) are output-order dependent, *i.e.*, under minimum variance control, $Var[y_1(t)]$ is minimized, $Var[y_2(t)]$ is also minimized but subject to the constraint that $Var[y_1(t)]$ is minimized, and so on. Therefore the importance of each output depends on the order it is stacked in the output vector, *i.e.*, the first output variable is the most important for the design of minimum variance control, and the last output variable is the least important. Re-arrangement of the output variables results in a different optimal control law. Shah *et al.* (1987) pointed out that the selection of the form of an interactor matrix is application-dependent, *i.e.*, it may take an upper triangular form or a full matrix form, and yet, in LRPC (long range predictive control) schemes for a specific choice of tuning parameters, the controller is independent of the interactor matrix. Rogozinski *et al.* (1987) proposed an algorithm for the factorization of the nilpotent interactor matrix which has the full-matrix form. Peng and Kinnaert (1992) found the existence of the unitary interactor matrix, which is a special form of the nilpotent interactor matrix. Since the unitary interactor is an all-pass term, the factorization of such a unitary interactor matrix does not change the spectral property of the underlying system, *i.e.*, the magnitude of the

24   4. Unitary Interactor Matrices and Minimum Variance Control

frequency response remains the same after the factorization. This property of the unitary interactor matrix is desirable for minimum variance control or singular LQ control and multivariate control loop performance assessment using minimum variance control as the benchmark. Here the term "singular LQ control" denotes LQ design without penalty to the control action.

The main contributions in this chapter are: 1) the extension of the unitary interactor matrix into the weighted unitary interactor matrix; 2) an alternative derivation of the optimal singular LQ or minimum variance control law with respect to the minimum variance control law (Goodwin and Sin 1984) and the singular LQ control law (Harris and MacGregor 1987); 3) proof of the equivalence of minimum variance control law (Goodwin and Sin 1984) and the singular LQ control law (Harris and MacGregor 1987) if a weighted unitary interactor matrix is used. This chapter is organized as follows: Section 4.2 introduces the unitary interactor matrix. In Section 4.3, the unitary interactor matrix is applied to the explicit solution of the minimum variance control law. The unitary interactor matrix is then extended to the weighted unitary interactor matrix, and the identity between the minimum variance control law and the singular LQ control law is established in Section 4.4. This is followed by concluding remarks in Section 4.6. Extension to the generalized unitary interactor matrix, which factors out both unstable (non-minimum phase) and infinite zeros and may be regarded as an alternative solution to the inner-outer factorization (Chu 1985), is discussed in Chapter 10. For the sake of simplicity in presentation, only the square transfer function matrix is considered in this chapter.

4.2 Unitary interactor matrices

The unitary interactor matrix has been defined in Chapter 3. The existence of the unitary interactor matrix is established in Peng and Kinnaert (1992).

Lemma 4.2.1. *For a full rank (in the field of q^{-1}) rational, proper transfer function matrix T, there exists a non-unique unitary interactor matrix. However, any two unitary interactor matrices, $D(q)$ and $\bar{D}(q)$, satisfy*

$$\bar{D}(q) = \Gamma D(q)$$

where Γ is an $n \times n$ unitary real matrix, i.e., $\Gamma^T \Gamma = I$.

Proof. See Peng and Kinnaert (1992) for the proof.

Readers are also referred to Peng and Kinnaert (1992) and Rogozinski et al. (1987) for the algorithm to factor the unitary interactor matrix from a transfer function matrix. This algorithm is summarized in Appendix A. In Chapter 5, this algorithm will be simplified by using QR decomposition of only the first few Markov parameter matrices or impulse response matrices.

a *priori* knowledge of the interactor matrix is tantamount to knowing the entire transfer function matrix, which is often a demanding requirement. One alternative is to simply compute the interactor matrix by estimating the first few Markov parameters of the closed-loop process via dither signal excitation.

The non-uniqueness of the interactor matrix can also be due to different ordering of the output variables, *i.e.*, the order in which output variables are stacked to form the output vector. The relationship between differently ordered unitary interactor matrices is established in the following lemma:

Lemma 4.2.2. *If $D(q)$ is the unitary interactor matrix of T, and $\bar{D}(q)$ is the unitary interactor matrix of the output-reordered transfer function matrix $\bar{T}(q^{-1}) = VT(q^{-1})$, where V is row-exchanging operator (an orthogonal matrix), then*

$$\bar{D}(q) = \Gamma D(q) V^T$$

where Γ is an $n \times n$ unitary real matrix.

Proof. From the definition of the unitary interactor matrix, we have

$$\lim_{q^{-1} \to 0} D(q) T(q^{-1}) = K_1 \tag{4.1}$$

$$\lim_{q^{-1} \to 0} \bar{D}(q) \bar{T}(q^{-1}) = \lim_{q^{-1} \to 0} \bar{D}(q) V T(q^{-1}) = K_2 \tag{4.2}$$

Recall that by definition, we also have $(D(q))^{-1} = D^T(q^{-1})$, $(\bar{D}(q))^{-1} = \bar{D}^T(q^{-1})$, $V^{-1} = V^T$.

From Equations 4.1 and 4.2, one can obtain

$$\lim_{q^{-1} \to 0} D(q) V^T \bar{D}^T(q^{-1}) = K_1 K_2^{-1} \triangleq \Gamma^{-1} \tag{4.3}$$

$$\lim_{q^{-1} \to 0} \bar{D}(q) V D^T(q^{-1}) = K_2 K_1^{-1} = \Gamma \tag{4.4}$$

It is obvious from Equations 4.3 and 4.4 that $\Gamma^{-1} = \Gamma^T$. Since $D(q) V^T \bar{D}^T(q^{-1})$ is a finite-order matrix polynomial, *i.e.*,

$$D(q) V^T \bar{D}^T(q^{-1}) \triangleq E(q, q^{-1})$$
$$= q^{-d+1} E_{-d+1} + \cdots + q^{-2} E_{-2} + q^{-1} E_{-1} + E_0 + q E_1 + q^2 E_2 + \cdots + q^{d-1} E_{d-1}$$

Equation 4.3 implies that $D(q) V^T \bar{D}^T(q^{-1})$ has no positive power of q. One may therefore write it as $D(q) V^T \bar{D}^T(q^{-1}) = E(q^{-1})$. On the other hand, Equation 4.4 also implies that $\bar{D}(q) V D^T(q^{-1}) = E^T(q)$ has no positive power of q, or equivalently, $E(q^{-1})$ has no negative power of q. Thus the matrix polynomial $E(q, q^{-1})$ is neither a function of q nor q^{-1}. It follows then from Equations 4.3 and 4.4 that

$$D(q) V^T \bar{D}^T(q^{-1}) = \Gamma^T$$
$$\bar{D}(q) V D^T(q^{-1}) = \Gamma$$

which yields
$$\bar{D}(q) = \Gamma D(q) V^T$$

Bittanti et al. (1994) have also defined a *spectral interactor matrix*, which has the same property as the right unitary interactor matrix defined by Panlinski and Rogozinski (1990). The unitary interactor matrix is an all-pass factor, as a delay-term should be, and retains the spectral property of the underlying system after the infinite zeros are removed and is an ideal factorization of time-delays for the design of minimum variance or singular LQ control. The advantage of factorizing a unitary interactor matrix as an all-pass factor is its computational simplicity compared to the spectral interactor factorization or the inner-outer factorization.

4.3 Unitary interactor matrices and the minimum variance control law

Goodwin and Sin (1984) have extended the deadbeat deterministic control strategy to the minimum variance control of systems with stable zeros. Consider a multivariable system
$$Y_t = T U_t + N a_t$$
where T is the system transfer function matrix and N is the disturbance transfer function matrix. The minimum variance control law can be designed to make the variance of the interactor-filtered output $D Y_t$ or equivalently $\tilde{Y}_t = q^{-d} D Y_t$ minimum, where the positive integer d is the order of the interactor matrix or the minimum integer which makes $q^{-d} D$ proper. This yields a simple multivariable control design strategy.

Theorem 4.3.1. *For a multivariable process*
$$Y_t = T U_t + N a_t \tag{4.5}$$
with the linear quadratic objective function (singular LQ objective function) defined by
$$J = E(\tilde{Y}_t^T \tilde{Y}_t) \tag{4.6}$$
where $\tilde{Y}_t = q^{-d} D Y_t$, an explicit optimal control law is given by
$$U_t = -\tilde{T}^{-1} R M_F^{-1} D Y_t = -\tilde{T}^{-1} R F^{-1} (q^{-d} D) Y_t \tag{4.7}$$
where $\tilde{T} = DT$, $M_F = q^d F$, F and R satisfy the identity:
$$q^{-d} D N = \underbrace{F_0 + \cdots + F_{d-1} q^{-d+1}}_{F} + q^{-d} R \tag{4.8}$$
and R is a rational proper transfer function matrix.

4.3 Unitary interactor matrices and the minimum variance control law

Proof. Consider the process with a general interactor matrix:

$$Y_t = TU_t + Na_t = D^{-1}\tilde{T}U_t + Na_t \quad (4.9)$$

Multiplying both sides of Equation 4.9 by $q^{-d}D$ yields

$$\begin{aligned} q^{-d}DY_t &= q^{-d}\tilde{T}U_t + q^{-d}DNa_t \\ &= q^{-d}\tilde{T}U_t + \tilde{N}a_t \end{aligned} \quad (4.10)$$

where $\tilde{N}$ is a proper transfer function matrix. By defining $\tilde{Y}_t = q^{-d}DY_t$, Equation 4.10 has been transformed to a process with a simple interactor matrix i.e.,

$$\tilde{Y}_t = q^{-d}\tilde{T}U_t + \tilde{N}a_t \quad (4.11)$$

Substituting Equation 4.8 into 4.11 yields

$$\tilde{Y}_t = \tilde{T}U_{t-d} + Ra_{t-d} + Fa_t \quad (4.12)$$

The last term in this equation cannot be affected by the control action, i.e.,

$$Var(\tilde{Y}_t) = E(\tilde{Y}_t\tilde{Y}_t^T) \geq Var(Fa_t)$$

Therefore

$$E(\tilde{Y}_t^T\tilde{Y}_t) \geq tr(Var(Fa_t))$$

The minimum variance control is achieved when the sum of the first two terms on the right hand side of Equation 4.12 is set to zero, i.e.,

$$\tilde{T}U_{t-d} + Ra_{t-d} = 0$$

This yields

$$U_t = -\tilde{T}^{-1}Ra_t \quad (4.13)$$

Substituting Equation 4.13 into 4.12 yields

$$\tilde{Y}_t = Fa_t \quad (4.14)$$

Therefore

$$a_t = F^{-1}\tilde{Y}_t \quad (4.15)$$

Substituting Equation 4.15 into 4.13 gives the minimum variance control law

$$U_t = -\tilde{T}^{-1}RF^{-1}\tilde{Y}_t = -\tilde{T}^{-1}RF^{-1}(q^{-d}D)Y_t \quad (4.16)$$

By defining $M_F = q^d F$, Equation 4.16 can be written as

$$U_t = -\tilde{T}^{-1}RM_F^{-1}DY_t$$

where F and R are defined by

$$q^{-d}DN = \underbrace{F_0 + \cdots + F_{d-1}q^{-d+1}}_{F} + q^{-d}R$$

or

$$DN = M_F + R$$

However, this minimum variance control law is only able to minimize variance of the interactor-filtered variable $\tilde{Y}_t$ as per the objective function defined by Equation 4.6. If D is a lower triangular interactor matrix as used by Goodwin and Sin (1984), then the minimum variance control law of $\tilde{Y}_t$ has the property that $Var[y_1(t)]$ is minimized, $Var[y_2(t)]$ is also minimized but subject to the constraint that $Var[y_1(t)]$ is minimized, and so on. Therefore the control law is output-order dependent (Dugard, Goodwin, and Xianya 1984). On the other hand, if D is a unitary interactor matrix, we have the following result:

Lemma 4.3.1. *If D is a unitary interactor matrix, then a proper optimal control law which minimizes the LQ objective function of the interactor-filtered output $\tilde{Y}_t$*

$$J_1 = E(\tilde{Y}_t^T \tilde{Y}_t) \tag{4.17}$$

also minimizes the LQ objective function of the original output Y_t

$$J_2 = E(Y_t^T Y_t) \tag{4.18}$$

and $J_1 = J_2$. Thus the singular LQ control law of the original variable Y_t can be obtained via the singular LQ control law of the unitary interactor-filtered variable $\tilde{Y}_t$.

Proof. Since a_t is random white noise with zero mean, we have

$$E(\tilde{Y}_t^T \tilde{Y}_t) = tr[Var(\tilde{Y}_t)]$$

Using Parseval's theorem (Ljung 1998) and noticing the property of the unitary interactor matrix, i.e.,

$$D^T(q^{-1})D(q) = I \quad \text{or} \quad D^T(e^{-j\omega})D(e^{j\omega}) = I \quad \text{(for all } \omega\text{)}$$

we have

$$\begin{aligned}
tr[Var(\tilde{Y}_t)] &= tr[Var(q^{-d}DY_t)] \\
&= \frac{1}{2\pi}\int_{-\pi}^{\pi} tr[D(e^{j\omega})\phi_Y(\omega)D^T(e^{-j\omega})]d\omega \\
&= \frac{1}{2\pi}\int_{-\pi}^{\pi} tr[D^T(e^{-j\omega})D(e^{j\omega})\phi_Y(\omega)] \\
&= \frac{1}{2\pi}\int_{-\pi}^{\pi} tr[\phi_Y(\omega)] \\
&= tr[Var(Y_t)] = E(Y_t^T Y_t)
\end{aligned}$$

where $\phi_Y(\omega)$ is the power spectrum density of Y_t, and the notation of the power spectrum density is given by (Ljung 1998): $\phi_Y(\omega) = \sum_{\tau=-\infty}^{\infty} R_Y(\tau)e^{-j\tau\omega}$ and $R_Y(\tau) = E(Y_t Y_{t-\tau}^T)$. Therefore $J_1 = J_2$, and the minimization of J_1 is equivalent to the minimization of J_2.

4.3 Unitary interactor matrices and the minimum variance control law

Another important property of the unitary interactor matrix for minimum variance control is that the control law is output-order invariant.

Lemma 4.3.2. *If D is a unitary interactor matrix, then the minimum variance control law as solved by Theorem 4.3.1 is output-ordering invariant.*

Proof. It follows from Theorem 4.3.1 that for the original system

$$Y_t = TU_t + Na_t$$

the minimum variance control law is

$$U_t = -\tilde{T}^{-1} R M_F^{-1} D$$

and for the output re-ordered system

$$\bar{Y}_t = VY_t = VTU_t + VNa_t = \bar{T}U_t + \bar{N}a_t$$

the minimum variance control law is

$$\bar{U}_t = -\tilde{\bar{T}}^{-1} \bar{R} \bar{M}_F^{-1} \bar{D} \bar{Y}_t \tag{4.19}$$

where $\bar{D}\bar{N} = \bar{M}_F + \bar{R}$ and $\tilde{\bar{T}} = \bar{D}\bar{T}$. From Lemma 4.2.2, we have

$$\bar{D} = \Gamma D V^T$$

Therefore

$$\tilde{\bar{T}} = \bar{D}\bar{T} = \Gamma D V^T V T = \Gamma D T = \Gamma \tilde{T}$$

and

$$\bar{M}_F + \bar{R} = \bar{D}\bar{N} = \Gamma D V^T V N = \Gamma D N = \Gamma(M_F + R)$$

Thus $\bar{M}_F = \Gamma M_F$ and $\bar{R} = \Gamma R$. Substituting $\tilde{\bar{T}}$, $\bar{R}$, $\bar{M}_F$, $\bar{D}$ and $\bar{Y}_t = VY_t$ into Equation 4.19 yields

$$\bar{U}_t = -\tilde{T}^{-1} \Gamma^T \Gamma R M_F^{-1} \Gamma^T \Gamma D V^T V Y_t = -\tilde{T}^{-1} R M_F^{-1} D Y_t = U_t$$

Lemma 4.3.3. *The minimum variance control law, as given in Theorem 4.3.1, is scaling invariant, i.e., if a interactor matrix D is pre-multiplied by an invertible constant matrix P ($\bar{D} = PD$), then using $\bar{D}$ as the interactor matrix results in the same control law as using D as the interactor matrix.*

Proof. For two interactor matrices, D and $\bar{D}$, with

$$\bar{D} = PD \tag{4.20}$$

we have

$$\tilde{\bar{T}} = \bar{D}T = \bar{D}D^{-1}\tilde{T} = PDD^{-1}\tilde{T} = P\tilde{T} \tag{4.21}$$

and

$$\bar{M}_F + \bar{R} = \bar{D}N = PDN = P(M_F + R) = PM_F + PR$$

Therefore

30   4. Unitary Interactor Matrices and Minimum Variance Control

$$\bar{M}_F = PM_F \qquad (4.22)$$
$$\bar{R} = PR \qquad (4.23)$$

The minimum variance controller(with $\bar{D}$ as its interactor matrix) is

$$\bar{U}_t = -\tilde{T}^{-1}\bar{R}\bar{M}_F^{-1}\bar{D}Y_t \qquad (4.24)$$

Substituting Equations 4.20, 4.21, 4.22 and 4.23 into Equation 4.24 yields

$$\bar{U}_t = -\tilde{T}^{-1}P^{-1}PRM_F^{-1}P^{-1}PDY_t = -\tilde{T}^{-1}RM_F^{-1}DY_t = U_t$$

Theorem 4.3.2. *If D is a unitary interactor matrix, the minimum variance control law as given by Theorem 4.3.1 is unique.*

Proof. Non-uniqueness of the unitary interactor matrix is due to output ordering and/or scaling, *i.e.*, $\bar{D} = \Gamma D$. It has been shown in Lemma 4.3.2 that the minimum variance control law is output order invariant. From Lemma 4.3.3, it follows that the unitary scaling matrix Γ does not affect the control law; therefore, the minimum variance control law with the unitary interactor matrix is unique.

4.4 Weighted unitary interactor matrices and singular LQ control

Definition 4.4.1. *In practice, individual outputs in a multivariate process often have different units, which may result in a significant differences in their magnitudes. In addition, each output may have different importance in the control objective. Thus, a weighted LQ objective function is often desired. The unitary interactor matrix, as introduced in the previous sections, is not able to solve weighted LQ or minimum variance control problem. It is, therefore, necessary to introduce alternative unitary interactor matrices.*

Instead of taking the lower triangular form or unitary interactor matrix form, if an interactor matrix as per Theorem 3.2.1 satisfies

$$D_w^T(q^{-1})D_w(q) = W \qquad (4.25)$$

where $W > 0$ is a symmetric weighting matrix, then this interactor matrix is regarded as the weighted unitary interactor matrix.

The weighted unitary interactor matrix has similar properties to the unitary interactor matrix. The existence of the weighted unitary interactor matrix is established in the following theorem:

4.4 Weighted unitary interactor matrices and singular LQ control

Theorem 4.4.1. *For a full rank (in the field of q^{-1}) rational, proper transfer function matrix T, there exists a non-unique weighted unitary interactor matrix. However, any two weighted unitary interactor matrices, $D_w(q)$ and $\bar{D}_w(q)$, satisfy*

$$\bar{D}_w(q) = \Gamma D_w(q)$$

where Γ is a $n \times n$ unitary real matrix, i.e., $\Gamma^T \Gamma = I$.

Proof. From the definition of the weighted unitary interactor matrix, we have

$$\lim_{q^{-1} \to 0} D_w(q) T(q^{-1}) = K_1 \qquad (4.26)$$

$$\lim_{q^{-1} \to 0} \bar{D}_w(q) T(q^{-1}) = K_2 \qquad (4.27)$$

From Equation 4.26 and 4.27, one can obtain

$$\lim_{q^{-1} \to 0} D_w(q)(\bar{D}_w(q))^{-1} = K_1 K_2^{-1} = \Gamma^{-1} \qquad (4.28)$$

$$\lim_{q^{-1} \to 0} \bar{D}_w(q)(D_w(q))^{-1} = K_2 K_1^{-1} = \Gamma \qquad (4.29)$$

From the definition (Equation 4.25), the following equations follow:

$$(D_w(q))^{-1} = W^{-1} D_w^T(q^{-1}) \qquad (4.30)$$
$$(\bar{D}_w(q))^{-1} = W^{-1} \bar{D}_w^T(q^{-1}) \qquad (4.31)$$

Substituting Equations 4.31 and 4.30 into 4.28 and 4.29 respectively yields

$$\lim_{q^{-1} \to 0} D_w(q) W^{-1} \bar{D}_w^T(q^{-1}) = K_1 K_2^{-1} = \Gamma^{-1} \qquad (4.32)$$

$$\lim_{q^{-1} \to 0} \bar{D}_w(q) W^{-1} D_w^T(q^{-1}) = K_2 K_1^{-1} = \Gamma \qquad (4.33)$$

It follows from Equations 4.32 and 4.33 that Γ is a unitary real matrix, i.e., $\Gamma^{-1} = \Gamma^T$. Equations 4.32 and 4.33 imply that $D_w(q) W^{-1} \bar{D}_w^T(q^{-1})$ and $\bar{D}_w(q) W^{-1} D_w^T(q^{-1})$ have neither a positive nor negative power of q. Therefore

$$D_w(q) W^{-1} \bar{D}_w^T(q^{-1}) = \Gamma^T$$
$$\bar{D}_w(q) W^{-1} D_w^T(q^{-1}) = \Gamma$$

It follows that

$$\bar{D}_w(q) = \Gamma D_w^{-T}(q^{-1}) W \qquad (4.34)$$

Substituting Equation 4.30 into 4.34 yields

$$\bar{D}_w(q) = \Gamma D_w(q)$$

Existence of the weighted unitary interactor matrix is given by Corollary 4.4.1.

32   4. Unitary Interactor Matrices and Minimum Variance Control

Corollary 4.4.1. *One of the solutions for the weighted unitary interactor matrix is given by*
$$D_w(q) = D(q)W^{1/2}$$
where $D(q)$ is a unitary interactor matrix of the weighted transfer matrix $W^{1/2}T(q^{-1})$. In general, any weighted unitary interactor matrix $D_w(q)$ can be written as
$$D_w(q) = \Gamma D(q)W^{1/2}$$

Proof. Since $D(q)$ is the unitary interactor matrix of $W^{1/2}T(q^{-1})$, $D_w(q) = D(q)W^{1/2}$ must be an interactor matrix of $T(q^{-1})$. Furthermore, from $D_w^T(q^{-1})D_w(q) = W^{1/2}D^T(q^{-1})D(q)W^{1/2} = W$, one can conclude that $D_w(q)$ is a weighted unitary interactor matrix. From Theorem 4.4.1, the general solution of the weighted unitary interactor matrix can be written as $D_w(q) = \Gamma D(q)W^{1/2}$.

Corollary 4.4.2. *If the interactor matrix is a weighted unitary interactor matrix D_w, the result obtained in Theorem 4.3.1 is equivalent to the solution of the weighted singular LQ control problem:*
$$J = E(Y_t^T W Y_t)$$
where W is the weighting matrix.

Proof. It follows from the same procedure as the proof of Lemma 4.3.1. Thus, the minimization of the variance of the interactor-filtered variable $\tilde{Y}_t$ by a proper optimal control law is equivalent to the minimization of the weighted variance of the original variable Y_t.

The unitary interactor matrix or weighted unitary interact matrix can be used for the design of singular LQ output feedback control law.

Theorem 4.4.2. *If a weighted unitary interactor is used for a process without non-invertible zeros, then its minimum variance control law as solved in Corollary 4.4.2 via Theorem 4.3.1 is the same as the singular LQ output feedback control law solved via spectral factorization (Harris and MacGregor 1987; Harris, Boudreau, and MacGregor 1996).*

Proof. For a MIMO process
$$Y_t = TU_t + Na_t$$
where N can be represented by an ARIMA model as $N = \Theta\Phi^{-1}$. Harris and MacGregor (1987) and Harris et al. (1996) have shown that the singular LQ control law (when input penalty matrix $Q_2 = 0$) is solved as
$$U_t = -H_1(Y_t - L\Lambda^{-1}U_t) = -H_1 N a_t = -H_1\Theta\Phi^{-1}a_t \qquad (4.35)$$
where $L\Lambda^{-1} = T$ is a matrix fraction representation of the transfer matrix T, and H_1 is a filter transfer matrix with

4.4 Weighted unitary interactor matrices and singular LQ control

$$H_1 = \tilde{T}^{-1} F_1 \tag{4.36}$$

where $\tilde{T}^{-1} = \Lambda \Gamma^{-1}$ is the optimal inverse of T (a proper inverse). Γ is solved from spectral factorization

$$\Gamma^H \Gamma = L^H W L \tag{4.37}$$

where W is the output weighting matrix [1] and the superscript H denotes the complex conjugate transpose, e.g., $\Gamma^H(q^{-1}) = \Gamma^T(q)$ and $L^H(q^{-1}) = L^T(q)$. F_1 is solved via

$$F_1 = \tau \Theta^{-1} \tag{4.38}$$

where τ is solved from the Diophantine equation

$$L^H W \Theta = \Gamma^H \tau + q P(q) \Phi \tag{4.39}$$

Left-multiplying both sides of Equation 4.37 by Λ^{-H} and right-multiplying by Λ^{-1}, and using the fact that $\tilde{T} = \Gamma \Lambda^{-1}$ and $T = L\Lambda^{-1}$, we have

$$\tilde{T}^H \tilde{T} = T^H W T \tag{4.40}$$

If a weighted unitary interactor is used as a factorization of time delays, then $\tilde{T} = D_w T$ does not contain any infinite zeros (time-delays), and Equation 4.40 or 4.37 is also satisfied, i.e., $\tilde{T}$ ($\tilde{T} = D_w T$) is the proper optimal inverse of T. Now left-multiplying Equation 4.39 by Λ^{-H} yields

$$T^H W \Theta = \tilde{T}^H \tau + \Lambda^{-H}[q P(q)] \Phi \tag{4.41}$$

Right-multiplying both sides of Equation 4.41 by Φ^{-1} yields

$$T^H W N = \tilde{T}^H \tau \Phi^{-1} + \Lambda^{-H}[q P(q)] \tag{4.42}$$

From the definition of the weighted unitary interactor, we have

$$T^H = (D_w^{-1} \tilde{T})^H = \tilde{T}^H D_w^{-H} = \tilde{T}^H D_w W^{-1}$$

Substituting this into Equation 4.42 yields

$$\tilde{T}^H D_w N = \tilde{T}^H \tau \Phi^{-1} + \Lambda^{-H}[q P(q)] \tag{4.43}$$

Multiplying Equation 4.43 by $\tilde{T}^{-H}$ results in

$$D_w N = \tau \Phi^{-1} + \tilde{T}^{-H} \Lambda^{-H}[q P(q)] = \tau \Phi^{-1} + \Gamma^{-H}[q P(q)] = R + M_F \tag{4.44}$$

The first term $R = \tau \Phi^{-1}$ is simply a proper matrix and involves only negative power terms of q^{-1}. The second term $M_F = \Gamma^{-H}[q P(q)] = \Gamma^{-T}(q)[q P(q)]$ involves only positive power terms of q. This has the same representation as in Theorem 4.3.1, and therefore M_F must be a finite order matrix polynomial.

Combining Equations 4.44, 4.38, 4.36 and 4.35 yields

---
[1] In Harris and MacGregor (1987), W is denoted as Q_1.

34   4. Unitary Interactor Matrices and Minimum Variance Control

$$U_t = -\tilde{T}^{-1}\tau\Phi^{-1}a_t = -\tilde{T}^{-1}Ra_t \tag{4.45}$$

Under this control law, the closed-loop response can be written as

$$\begin{aligned}
Y_t &= TU_t + Na_t \\
&= -D_w^{-1}\tilde{T}\tilde{T}^{-1}Ra_t + D_w^{-1}D_wNa_t \\
&= -D_w^{-1}Ra_t + D_w^{-1}[M_F + R]a_t \\
&= D_w^{-1}M_Fa_t
\end{aligned} \tag{4.46}$$

Therefore

$$a_t = M_F^{-1}D_wY_t$$

Substituting this into Equation 4.45 yields

$$U_t = -\tilde{T}^{-1}RM_F^{-1}D_wY_t \tag{4.47}$$

This yields the same control law as Theorem 4.3.1 with the weighted unitary interactor matrix, D_w.

4.5 Numerical Example

Consider a 2×2 multivariable process, with the open-loop transfer function matrix T and disturbance transfer function matrix N given by

$$T = \begin{bmatrix} \frac{q^{-1}}{1-0.4q^{-1}} & \frac{K_{12}q^{-2}}{1-0.1q^{-1}} \\ \frac{0.3q^{-1}}{1-0.1q^{-1}} & \frac{q^{-2}}{1-0.8q^{-1}} \end{bmatrix}$$

$$N = \begin{bmatrix} \frac{1}{1-0.5q^{-1}} & \frac{-0.6}{1-0.5q^{-1}} \\ \frac{0.5}{1-0.5q^{-1}} & \frac{1.0}{1-0.5q^{-1}} \end{bmatrix}$$

where K_{12} is any constant. Suppose that the LQ objective function is given by

$$J = E[Y_t^T Y_t]$$

Then, a unitary interactor matrix is required for the design of the optimal control law. Following the procedure in Appendix A, a unitary interactor matrix D can be factored as

$$D = \begin{bmatrix} -0.9578q & -0.2873q \\ -0.2873q^2 & 0.9578q^2 \end{bmatrix}$$

and the order of the interactor matrix $d = 2$. Thus, DN can be calculated as

$$DN = \begin{bmatrix} \frac{-1.1014q}{(1-0.5q^{-1})} & \frac{0.2874q}{(1-0.5q^{-1})} \\ \frac{0.1916q^2}{(1-0.5q^{-1})} & \frac{1.1302q^2}{(1-0.5q^{-1})} \end{bmatrix}$$

From $q^{-d}DN = F + q^{-d}R$, one can calculate F and R as

4.5 Numerical Example

$$F = \begin{bmatrix} -1.1014q^{-1} & 0.2874q^{-1} \\ 0.1916 + 0.0958q^{-1} & 1.1302 + 0.5651q^{-1} \end{bmatrix}$$

$$R = \begin{bmatrix} \frac{-0.5507}{1-0.5q^{-1}} & \frac{0.1437}{1-0.5q^{-1}} \\ \frac{0.0479}{1-0.5q^{-1}} & \frac{0.2826}{1-0.5q^{-1}} \end{bmatrix}$$

The optimal (minimum variance) control law can then be calculated from Equation 4.7. The interactor-filter output ($\tilde{Y}_t = q^{-d}DY_t$) under optimal control is given by Equation 4.14:

$$\tilde{Y}_t|_{mv} = Fa_t = \begin{bmatrix} -1.1014q^{-1} & 0.2874q^{-1} \\ 0.1916 + 0.0958q^{-1} & 1.1302 + 0.5651q^{-1} \end{bmatrix} a_t$$

Now consider a weighted LQ objective

$$J = E[Y_t^T W Y_t]$$

Suppose the weighting matrix is given by

$$W = \begin{bmatrix} 1 & 0 \\ 0 & 4 \end{bmatrix}$$

It follows from Corollary 4.4.1 that the weighted interactor matrix is given by

$$D_w = DW^{1/2}$$

where D is a unitary interactor matrix of the weighted transfer function matrix $W^{1/2}T$. Following the procedure in Appendix A, the unitary interactor matrix D is calculated as

$$D = \begin{bmatrix} -0.8575q & -0.5145q \\ 0.5145q^2 & -0.8575q^2 \end{bmatrix} \qquad (4.48)$$

Thus, the weighted unitary interactor matrix is

$$D_w = \begin{bmatrix} -0.8575q & -1.029q \\ 0.5145q^2 & -1.715q^2 \end{bmatrix}$$

The matrices F and R can be calculated from $q^{-d}D_w N = F + q^{-d}R$ as

$$F = \begin{bmatrix} -1.3720q^{-1} & -0.5145q^{-1} \\ -0.3430 - 0.1715q^{-1} & -2.0240 - 1.0120q^{-1} \end{bmatrix}$$

$$R = \begin{bmatrix} \frac{-0.6860}{1-0.5q^{-1}} & \frac{-0.2572}{1-0.5q^{-1}} \\ \frac{-0.0858}{1-0.5q^{-1}} & \frac{-0.5060}{1-0.5q^{-1}} \end{bmatrix}$$

The optimal (minimum variance) control law can then be calculated from Equation 4.7 with the interactor matrix D substituted by the weighted unitary interactor matrix D_w, i.e.,

$$U_t = -\tilde{T}^{-1}RM_F^{-1}D_w Y_t = -(D_w T)^{-1}RF^{-1}(q^{-d}D_w)Y_t$$

The interactor-filtered output ($\tilde{Y}_t = q^{-d}D_w Y_t$) under optimal control is given by Equation 4.14 as

$$\tilde{Y}_t|_{mv} = Fa_t = \begin{bmatrix} -1.3720q^{-1} & -0.5145q^{-1} \\ -0.3430 - 0.1715q^{-1} & -2.0240 - 1.0120q^{-1} \end{bmatrix} a_t$$

4.6 Summary

This chapter has shown that the unitary/weighted-unitary interactor matrix is an "ideal" factorization of the time-delays of multivariable systems for the design of minimum variance control or singular LQ control. It gives a unique minimum variance control that is output-ordering invariant and also scaling invariant. Using the unitary/weighted-unitary interactor matrix, the simple multivariable minimum variance control strategy as proposed by Goodwin and Sin (1984) gives a unique solution which is identical to the singular LQ output feedback control law (Harris and MacGregor 1987). This result is particularly useful for multivariable control loop performance assessment and for the design of singular LQ control of a minimum phase MIMO process.

CHAPTER 5
ESTIMATION OF THE UNITARY INTERACTOR MATRICES

5.1 Introduction

The idea of the time delay term can be easily generalized to the multivariate case in terms of the impulse response coefficient or the Markov parameter matrices. In the multivariate case, the notion of a delay corresponds to the fewest number of impulse response coefficients or Markov parameter matrices whose linear combination is *nonsingular*. This means that a set of inputs, acting via this specific linear combination of Markov parameter matrices, can have a desired effect on the output. This linear combination of impulse response matrices can be expressed in a polynomial matrix form. The determinant of this polynomial matrix has, as its roots, the infinite zeros of the discrete time multivariate system. Simple examples to illustrate these concepts are considered in Shah *et al.* (1987). The knowledge of the interactor matrix is an important prerequisite to high performance control strategies such as minimum variance control. However a knowledge of the delay or the interactor matrix is, until recently, tantamount to the knowledge of the entire process transfer function matrix. As per the above definition, it should appear that relatively simple tests can be performed to determine if a linear combination of the first few Markov or impulse response matrices is singular or not. This is precisely the purpose of this chapter in which we propose the use of a singular value decomposition (SVD) based procedure to determine if a linear combination of a set of matrices has full rank. Note that the determination of a rank of a matrix is a non-trivial computational problem. SVD-based techniques are useful in determining the rank of a given matrix. The proposed procedure allows us to compute the time delay matrix with minimum effort using closed- or open-loop data with dither excitation and its subsequent use in multivariate control loop performance assessment or control law design.

The algorithm for factoring the lower triangular interactor matrix, as suggested by Wolovich and Falb (1976) and Goodwin and Sin (1984), generally requires a complete knowledge of the transfer function matrix. Shah *et al.* (1987) and Mutoh and Ortega (1993), however, have suggested a solution of the interactor matrix by solving a set of linear, algebraic equations of certain Markov parameter matrices (impulse response coefficient matrices). This latter approach connects the Markov parameter matrices to the interactor matrix directly without going through the transfer function and is numerically convenient and attractive for estimation of the interactor matrix of a MIMO process. The lower triangular interactor matrix has played an important role in classical multivariable control. Readers are referred to Walgama (1986) and Sripada (1988) for interesting discussions on this issue. Shah et al. (1987) and Rogozinski et al. (1987) have also pointed out that the interactor matrix need not necessarily take the lower-triangular form in application: for example, an interactor matrix with a unit-DC gain or other useful features may be more important in practice, and therefore the interactor matrix (as proposed by Wolovich and Falb (1976)) is not necessarily unique in the sense that it can have forms other than the lower triangular form. However, the "optimal" form of the interactor matrix is application dependent. As pointed out earlier, for a deterministic system, optimal control design based on the lower triangular interactor matrix yields a conditional minimum-time or minimum-ISE control in the sense that the optimization is input-output-pairing- or order-dependent (Tsiligiannis and Svoronos 1988). For a stochastic system, optimal control design based on the lower triangular interactor matrix yields a conditional minimum variance control (Dugard, Goodwin, and Xianya 1984). Peng and Kinnaert (1992) and Bittanti et al. (1994) have introduced the *unitary* or *spectrum* interactor matrix for the design of a singular LQ state feedback controller and optimal filter. The unitary interactor matrix simplifies the design procedure of the singular LQ control law. It gives an alternative derivation of the LQ controller for processes without finite unstable zeros and with output penalty matrix $Q_1 = I$ and control weighting $Q_2 = 0$, *i.e.*, $J = E[Y^T Y]$. The unitary interactor matrix can be easily extended to the *weighted unitary interactor matrix*. This weighted unitary interactor matrix can then be used for the design of the weighted singular LQ controller, *i.e.*, a controller which minimizes $J = E[Y^T Q_1 Y]$. The unitary interactor matrix is in fact a special case of the nilpotent interactor matrix as defined by Rogozinski and co-workers (1987, 1990), and plays an important role in multivariate control loop performance assessment theory.

For closed-loop control loop performance assessment, the estimation of the interactor matrix under closed-loop conditions is desired. In this chapter, an algorithm for the estimation of the unitary interactor matrix is proposed. Using the proposed method, the interactor matrix can be estimated from closed-loop data without estimation of the open-loop transfer function matrix.

The main contributions of this chapter are as follows: 1) development of a new method for determination of the order of the interactor matrix by using the singular value decomposition technique; 2) the extension of the results in Rogozinski et al. (1987) and Peng and Kinnaert (1992) for factorization of the unitary interactor by using the first few Markov parameters of a transfer function matrix; 3) the use of closed-loop data for the estimation of the Markov parameters of the transfer function matrix; 4) the experimental evaluation and industrial application of the proposed algorithm. Unlike other interactor factorization methods, which generally require complete knowledge of the entire transfer function matrix, this algorithm only requires the first few Markov parameter matrices.

This chapter is organized as follows: The method for determination of the order of the interactor matrix is developed in Section 5.2. The algorithm for the calculation of the unitary interactor matrix is then introduced in Section 5.3. The estimation of the unitary interactor matrix under closed-loop conditions is given a detailed treatment in Section 5.4. The determination of a numerical rank is discussed in Section 5.5. The chapter ends with an illustration with a simulated example, an experimental-data based application on a pilot-scale experiment in Section 5.6, and an industrial application in Section 5.7.

5.2 Determination of the order of interactor matrices

The interactor matrix has been given in Theorem 3.2.1. Wolovich and Falb (1976), and Goodwin and Sin (1984) have suggested factoring a lower triangular interactor matrix from the transfer function matrix. To do this, *a priori* knowledge of the entire transfer matrix is generally required. This is a fairly strong requirement. Shah et al. (1987) have suggested factoring the interactor matrix directly from Markov parameters of the process. This idea is further explored for the determination of the order of the interactor matrix.

The Markov parameter representation of a transfer function matrix can be written as

$$T = \sum_{i=0}^{\infty} G_i q^{-i-1} \qquad (5.1)$$

and the interactor matrix is written as

$$D = D_0 q^d + D_1 q^{d-1} + \cdots + D_{d-1} q \qquad (5.2)$$

From Theorem 3.2.1

$$\lim_{q^{-1} \to 0} DT = \lim_{q^{-1} \to 0} [D_0 q^d + D_1 q^{d-1} + \cdots + D_{d-1} q][G_0 q^{-1} + G_1 q^{-2} + \cdots] = K$$

where K is a full rank matrix (*i.e.*, rank(K)=min(n,m)), we have

$$D_0 G_0 = 0$$
$$D_1 G_0 + D_0 G_1 = 0$$
$$\vdots$$
$$D_{d-1} G_0 + \cdots + D_1 G_{d-2} + D_0 G_{d-1} = K$$

Solving the above algebraic equations yields the general solution of the interactor matrix.

5. Estimation of the Unitary Interactor Matrices

The above algebraic equations can be further written in a matrix form as

$$[D_{d-1}, \cdots, D_0] \begin{bmatrix} G_0 & 0 & 0 & \cdots & 0 \\ G_1 & G_0 & 0 & \cdots & 0 \\ \vdots & \vdots & \ddots & \ddots & \vdots \\ G_{d-2} & G_{d-3} & \cdots & \ddots & 0 \\ G_{d-1} & G_{d-2} & \cdots & \cdots & G_0 \end{bmatrix} = [K, 0, \cdots, 0] \quad (5.3)$$

or for simplicity

$$\underline{D}\,\underline{G} = \underline{K} \quad (5.4)$$

where $\underline{G}$ is a block-Toeplitz matrix. $\underline{D}$ denotes the algebraic matrix form of the interactor matrix while D is the matrix polynomial form of the interactor matrix, and $\underline{K} = [K, 0, \cdots, 0]$. If G_0 is not full rank, then in addition to the infinite zeros owing to the zero-order-hold, at least one more infinite zero exists in the transfer function matrix. Direct inversion for solving Equation 5.4 is impossible owing to $\underline{G}$ being rank deficient. The existence of the solution for Equation 5.4 also depends on the order of the interactor matrix d, i.e., the "size" of $\underline{G}$ such that there is at least an exact solution of $\underline{D}$. For determining the order of the interactor matrix, the singular value decomposition technique can be used.

Consider the singular value decomposition[1] of the block-Toeplitz matrix as

$$\underline{G} = U\Sigma V^T = [U_1, U_2] \begin{bmatrix} \Sigma_r & 0 \\ 0 & 0 \end{bmatrix} \begin{bmatrix} V_1^T \\ V_2^T \end{bmatrix} \quad (5.5)$$

where $[U_1, U_2]$ and $[V_1, V_2]^T$ are orthogonal matrices, the columns of U_2 span the null space of $\underline{G}$ (in the sense that $U_2^T \underline{G} = 0$), Σ_r is a full rank diagonal matrix, and the rows of V_1^T span the *row* space of $\underline{G}$.

The existence of the exact solution for Equation 5.4 requires that 1) rank($\underline{G}$) $\geq$ rank($\underline{K}$)=rank(K)=min(n,m); and 2) each row of $\underline{K}$ must be within the row space spanned by V_1^T or orthogonal to the row space spanned by V_2^T, i.e.,

$$\underline{K} V_2 = 0 \quad (5.6)$$

This can be simplified by writing

$$\underline{K} V_2 = [K, 0, \cdots, 0] \begin{bmatrix} V_{21} \\ V_{22} \\ \vdots \\ V_{2d} \end{bmatrix} = K V_{21} \quad (5.7)$$

---

[1] Note here that the linear matrix equation is in the form of $XA = B$, instead of $AX = B$, where X is the unknown vector or matrix. The definitions of the null space and the image space of A for the two equations are consequently different.

where V_{21} is the upper partition of V_2 with its row dimension the same as the column dimension of T. Thus, the condition expressed by Equation 5.6 is equivalent to

$$KV_{21} = 0 \qquad (5.8)$$

If K (or T) is a square matrix or is an $n \times m$ non-square matrix with $n > m$, Equation 5.8 is further simplified to

$$V_{21} = 0 \qquad (5.9)$$

If, however, these conditions are not satisfied, the block-Toeplitz matrix must be expanded by adding more Markov parameters until they are satisfied. Thus, the order of the interactor matrix d can be determined from Equation 5.8 or 5.9.

If T is a square transfer function matrix, then the *nullity increasing property* of the block-Toeplitz matrix (see Remark 5.2.1) can also be used conveniently to determine the order of the interactor matrix.

Remark 5.2.1. Mutoh and Ortega (1993) have suggested using the *nullity increasing property* of Markov parameters for determination of the order of the interactor matrix, d, of a square transfer function matrix. According to the nullity increasing property, the dimension of null space of $\underline{G}$ increases with expansion of $\underline{G}$ until all d Markov parameters are included in Matrix $\underline{G}$.

In summary, the result presented in this section is useful not only for the factorization of the interactor matrix, as discussed in the following sections, but also in the design of multivariate adaptive control without a complete knowledge of the interactor matrix (Shah, Mohtadi, and Clarke 1987).

5.3 Factorization of unitary interactor matrices

The solution of Equation 5.4 is not unique. The "optimal" solution depends on the application. The unitary interactor matrix discussed in this section is one of several such "optimal" solutions for the application in minimum variance control and multivariable control loop performance assessment. Rogozinski *et al.* (1987) have introduced the *nilpotent interactor matrix*. For a class of interactor matrices which are more suitable for LQ design, Peng and Kinnaert (1992) have further considered the *unitary interactor matrix*, which is a special case of the nilpotent interactor matrix. Bittanti *et al.* (1994) have also defined a *spectral interactor matrix*, which has essentially the same property as the *right unitary interactor matrix* discussed in a separate paper by Panlinski and Rogozinski (1990). In Chapter 4, the unitary interactor matrix has been shown to be a suitable factorization of the time delay for minimum variance or singular LQ control. It keeps the spectral property of the underlying system unchanged after the infinite zeros of the transfer matrix are removed. The "Inner-Outer" factorization as introduced in (Chu 1985) factors out an "all-pass" transfer matrix which also maintains the spectral property, but it requires the solution of an algebraic Riccati equation. Significant additional effort and process information are then required to factor out the infinite zeros from the "Inner-Outer factorization".

The algorithm for the calculation of the unitary interactor matrix proposed by Rogozinski et al. (1987) requires the right matrix fraction (RMF) of the transfer matrix. This is tantamount to knowing the entire transfer function matrix. In the present chapter, if the Markov parameter representation is used, the algorithm can be simplified.

Assumption 1. T is of a full rank $n \times m$ rational polynomial transfer function matrix, i.e., $rank[T(q^{-1})] = min(n, m)$.

Assumption 2. T is proper, i.e., $\lim_{q^{-1} \to 0} T(q^{-1}) < \infty$.

A block matrix of the first d Markov parameters is expressed in a block matrix form as

$$\Lambda = [G_0^T, G_1^T, \cdots G_d^T]^T$$

Once this block matrix Λ is formed, the unitary interactor matrix $D(q)$ can be factored out from this block matrix following the procedure in Rogozinski et al. (1987), and Peng and Kinnaert (1992). However, the numerator matrix coefficients of the right matrix fraction (RMF) of T would be replaced by the first d Markov parameter matrices. Note that even without the knowledge of the interactor matrix order, d, the algorithm can also factor the unitary interactor matrix but must include enough Markov parameter matrices into Λ by trial and error, for example.

Example 5.3.1. A numerical example is given to illustrate the proposed algorithm

Consider a 2×2 transfer function matrix:

$$T = \begin{bmatrix} \frac{q^{-1}}{1-0.1q^{-1}} & \frac{q^{-1}}{1-0.1q^{-1}} \\ \frac{2q^{-1}}{1-0.3q^{-1}} & \frac{2q^{-1}}{1-0.4q^{-1}} \end{bmatrix}$$

$$= \begin{bmatrix} 1 & 1 \\ 2 & 2 \end{bmatrix} q^{-1} + \begin{bmatrix} 0.1 & 0.1 \\ 0.6 & 0.8 \end{bmatrix} q^{-2} + \cdots$$

$$\triangleq G_0 q^{-1} + G_1 q^{-2} + \cdots \tag{5.10}$$

To determine the order of the interactor matrix, using the SVD method (skipping the first step $\underline{G} = G_0$ which is obviously rank deficient) gives

$$\underline{G} = \begin{bmatrix} G_0 & 0 \\ G_1 & G_0 \end{bmatrix} = \begin{bmatrix} 1 & 1 & 0 & 0 \\ 2 & 2 & 0 & 0 \\ 0.1 & 0.1 & 1 & 1 \\ 0.6 & 0.8 & 2 & 2 \end{bmatrix} \tag{5.11}$$

Using the SVD decomposition ($\underline{G} = U\Sigma V^T$) and Equation 5.5 gives

$$U_2 = \begin{bmatrix} 0.8944 \\ -0.4472 \\ 0 \\ 0 \end{bmatrix} \quad V_2 = \begin{bmatrix} 0 \\ 0 \\ -0.7071 \\ 0.7071 \end{bmatrix} \quad \Sigma_r = \begin{bmatrix} 3.6812 & 0 & 0 \\ 0 & 2.7322 & 0 \\ 0 & 0 & 0.0629 \end{bmatrix}$$

The column dimension of T is 2, so the upper partition of V_2 can be written as

$$V_{21} = \begin{bmatrix} 0 \\ 0 \end{bmatrix}$$

and $rank(\underline{G}) > rank(K) = min(n,m) = 2$; therefore, the conditions for the existence of the interactor matrix are satisfied, and consequently, the order of the interactor matrix $d = 2$ is selected. The block matrix of the first two Markov parameters can be formed as

$$\Lambda = \begin{bmatrix} 1 & 2 & 0.1 & 0.6 \\ 1 & 2 & 0.1 & 0.8 \end{bmatrix}^T$$

Following the algorithm in Rogozinski et al. (1987) (also see Appendix A), a unitary interactor matrix can be factored as

$$D = \begin{bmatrix} -0.4472q & -0.8944q \\ -0.8944q^2 & 0.4472q^2 \end{bmatrix} \tag{5.12}$$

It can be easily verified that $D^T(q^{-1})D(q) = I$.

5.4 Estimation of the interactor matrix under closed-loop conditions

The proposed factorization algorithm requires only the first d Markov parameter matrices (or impulse response coefficient matrices), i.e., the first several steps of the initial responses of a system. Since the first few Markov parameter matrices contribute to the initial transient response of the process, these parameters characterize the high frequency dynamics of the process. Thus, the interactor matrix, which consists of a linear combination of the first few Markov parameters, typically represents the high-frequency gain (Shah, Mohtadi, and Clarke 1987) of a system. An identification strategy which can yield good estimates in the high frequency range is more desired. A relatively high-frequency dither signal may be used for such a purpose. Computationally, a correlation analysis generally provides a relatively good estimate of these first few Markov parameters, since $Var(\hat{G}_k) \propto 1/(N-k)$(Box and Jenkins 1976), where $\hat{G}_k$ is the estimated Markov parameter via cross correlation analysis, and N is total number of data points used for the estimation. Alternatively, a parametric model can also be fitted from input-output data,

5. Estimation of the Unitary Interactor Matrices

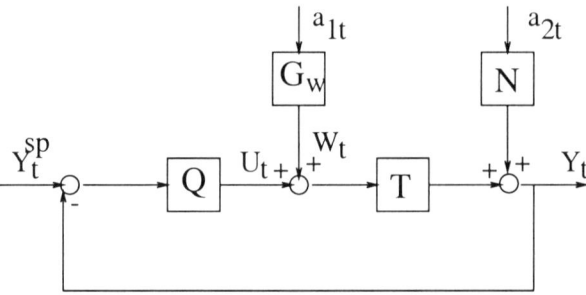

Fig. 5.1. A simplified process control loop diagram

which can also be transferred into the Markov parameters form. By utilizing the following lemma, correlation analysis or parametric model fitting can be performed directly from closed-loop data.

Lemma 5.4.1. *For a multivariable process as shown in Figure 5.1, the interactor matrix (D_{cl}) of the closed-loop transfer function matrix from W_t to Y_t ($T_{cl} = (I + TQ)^{-1}T$) is the same as the interactor matrix (D) of the open-loop transfer function matrix (T)*

Proof. It follows from the matrix inversion lemma (Soderstrom and Stoica 1989) that

$$\begin{aligned} T_{cl} &= (I + TQ)^{-1}T \\ &= [I - T(QT + I)^{-1}Q]T \\ &= T(I + QT)^{-1} \end{aligned}$$

Thus, if D is the interactor matrix of T, then $\lim_{q^{-1} \to 0} DT = K$ and

$$\begin{aligned} \lim_{q^{-1} \to 0} DT_{cl} &= \lim_{q^{-1} \to 0} DT(I + QT)^{-1} \\ &= \lim_{q^{-1} \to 0} DT(I + 0)^{-1} \text{ (due to the zero-order-hold)} \\ &= K \end{aligned}$$

On the other hand, if D_{cl} is the interactor matrix of T_{cl}, then $\lim_{q^{-1} \to 0} D_{cl}T_{cl} = K_{cl}$, or

$$\lim_{q^{-1} \to 0} D_{cl}T(I + QT)^{-1} = K_{cl}$$

Thus

$$\lim_{q^{-1} \to 0} D_{cl}T = K_{cl}$$

and, therefore, D_{cl} is also the interactor matrix of the open-loop transfer function matrix T.

5.4 Estimation of the interactor matrix under closed-loop conditions

If the dither signal is inserted from the setpoint, the same conclusion holds for the closed-loop transfer function matrix from Y_t^{sp} to Y_t following the same procedure of the proof, provided that the controller transfer function matrix does not introduce new infinite zeros to the process.

Remark 5.4.1. This lemma demonstrates the well-known fact that the delay structure or the interactor matrix is "feedback control invariant"(Wolovich and Falb 1976), *i.e.*, the Markov parameters of the open and closed-loop transfer function matrix are different, but their linear combination yields the same interactor matrix. With this result, the interactor matrix of an open-loop transfer function can be estimated directly from the closed-loop data.

Whenever the dither signal is "white" or can be whitened by time series analysis, simple correlation analysis can be performed. For actual plants, a random dither signal may not be allowed. For such cases, a series of simple step changes of the setpoint may be conducted instead, and parametric models may be fitted from data. As mentioned earlier, however, one must keep in mind that a low-frequency dither signal may yield a poorer estimate of the interactor matrix than a relatively high-frequency dither signal.

Our purpose is to identify the interactor matrix from closed-loop data via the dither signal or the setpoint to the output. A MISO identification procedure may be used if the dither signals or setpoint changes of all loops are conducted simultaneously. If, however, the dither signal or setpoint change of each loop is conducted separately, a SISO identification procedure may also be used. In addition, incorporate all *a priori* knowledge of the process, such as time delays, into the models. This will reduce mistakes in determining the rank of the Markov matrices.

Remark 5.4.2. A typical industrial process could be of very high order and possibly non-linear. A process under regulatory control usually operates around a nominal point. Identification of the interactor matrix under closed-loop conditions therefore provides a more realistic estimate than under open-loop conditions in the sense that it gives the interactor matrix of the process around the current operating point. This property is particularly useful for adaptive control and control loop performance monitoring. Similarly, a good estimate of the first few Markov parameters or initial transient responses is more important than a good estimate of the overall transfer function matrix, which often makes a compromise fit over a wider frequency range. Computationally, a direct identification of the first few Markov parameters is also more desirable than an identification of the full transfer function matrix first, followed by its transfer to Markov parameters. Therefore, the factorization of the interactor matrix from the first few Markov parameters is preferred to the factorization of the interactor matrix from the transfer function matrix.

5.5 Numerical rank

The estimated Markov parameter matrices are not exact owing to disturbances, and this makes numerical determination of the rank of the block-Toeplitz matrix $\underline{G}$ somewhat arbitrary. To cope with the difficulty, a result from Aoki (1987) (also see Paige (1981)) is used.

Let H be a theoretical matrix with its theoretically exact singular value decomposition, $U \Sigma V^T$. Suppose that a numerically constructed approximation to H is available as $\hat{H} = H + \Delta H$, where it is known that $||H - \hat{H}|| \leq a||\hat{H}||$. Here, the constant a represents a measure of data accuracy. If the computer round-off error is omitted, then in terms of the singular values of H and $\hat{H}$, this inequality can be stated as

$$|\sigma_i - \hat{\sigma}_i| \leq a\hat{\sigma}_1$$

where $\sigma_i, \hat{\sigma}_i$ denote the ith theoretical singular and calculated singular values respectively. If $\hat{\sigma}_r$ is greater than $a\hat{\sigma}_1$, but $\hat{\sigma}_{r+1}$ is less than this number, then clearly σ_r is positive. Hence the rank of the matrix is at least r. The next singular value, σ_{r+1}, may be zero. Such an r may then be chosen as the numerical rank of the true but unknown matrix H.

Golub and van Loan(1989) have another useful result in which the difference of the singular value of $H + E$ and H is bounded by the largest singular value of E, where E is considered as a perturbation matrix. We will regard this largest singular value as the threshold value.

The above results can be applied to find the rank of G_i and $\underline{G}$. The Aoki approach requires *a priori* knowledge of a. To find the value a, we may use an empirical value or a statistical value. As an example, Tiao and Box (1981) use $2/\sqrt{M}$ (where M is the sample size) as the relative error, a.

The threshold value approach is also useful if some pre-knowledge of perturbation is available. The correlation analysis or FIR model fitting often provides such knowledge. In addition to the Markov parameter matrices expressed by Equation 5.1, the Markov parameter matrix corresponding to the zero order of q, written as G_{-1} in accordance with Equation 5.1, is also obtained simultaneously in the correlation analysis. Owing to zero-order-hold, G_{-1} is, theoretically, zero, and, therefore, it does not appear in Equation 5.1. Its estimation is not zero owing to the effect of disturbances, however. Thus, the estimated value of G_{-1} provides an approximation of the perturbation matrix, E, and can be used to determine the rank of the estimated Markov parameters. Once the order of the interactor matrix is determined, the column block matrix of the Markov parameters can be formed, and the factorization of the unitary interactor matrix can proceed.

5.6 Simulation and experimental evaluation on a pilot scale process

Example 5.6.1. A closed-loop multivariable process, represented by the block diagram shown in Figure 5.1, is simulated. The interactor factorization algorithm based on correlation analysis is used to find the closed-loop interactor matrix, and the results are compared with the open-loop interactor matrix. To keep routine operation of process and to show the asymptotic property of correlation analysis, the magnitude of the dither signal is chosen such that it has a very weak effect on the process output relative to the existing process disturbances.

For the sake of comparison, we use the same open-loop transfer function T as the one used in Example 5.3.1. The remaining transfer function matrices of Figure 5.1 take the following values:

$$Q = \begin{bmatrix} 0.4 & 0 \\ 0 & 0.3 \end{bmatrix} \quad N = \begin{bmatrix} 1 & 2 \\ 3 & 4 \end{bmatrix} \quad G_w = \begin{bmatrix} 1 & 0 \\ 0 & 1 \end{bmatrix}$$

The setpoint is assumed to be zero. a_{1t} and a_{2t} are white noise random processes with $\Sigma_{a1} = 0.07^2 I$ and $\Sigma_{a2} = 0.1^2 I$. In this simulation, the existing variance of the output without the dither signal has a magnitude of

$$\Sigma_Y = \begin{bmatrix} 0.2203 & 0.3958 \\ 0.3958 & 0.7402 \end{bmatrix} \quad (5.13)$$

With injection of the dither signal, variance of the process output becomes

$$\Sigma_Y = \begin{bmatrix} 0.2381 & 0.4268 \\ 0.4268 & 0.7956 \end{bmatrix}$$

Thus, the dither signal has a negligible effect on the process. Since the dither signal is weak, a relatively large sample size of 5000 points is used[2]. Applying the cross-correlation analysis, the first three Markov parameter matrices (including G_{-1}) are calculated as

$$\hat{G}_{-1} = \begin{bmatrix} -0.0223 & 0.0135 \\ -0.0268 & -0.0182 \end{bmatrix} \quad \hat{G}_0 = \begin{bmatrix} 1.0161 & 0.9824 \\ 2.0435 & 1.9626 \end{bmatrix}$$

$$\hat{G}_1 = \begin{bmatrix} -0.9321 & -0.8885 \\ -1.4394 & -1.1543 \end{bmatrix}$$

The SVD decomposition of $\hat{G}_{-1}$ yields the largest singular value as 0.0355. Thus, the value of 0.0355 can be taken as the threshold to decide if a singular value is significantly different from zero. We can also use the relative error $a = 2/\sqrt{N} = 2/\sqrt{5000} = 0.0283$ to test the rank.

---

[2] The sample size can certainly be reduced if the magnitude of the dither signal is increased.

48   5. Estimation of the Unitary Interactor Matrices

Form the matrix $\underline{G}$ as $\underline{G} = \hat{G}_0$. The SVD decomposition of $\underline{G}$ yields

$$U = \begin{bmatrix} 0.4464 & 0.8948 \\ 0.8948 & -0.4464 \end{bmatrix} \quad \Sigma = \begin{bmatrix} 3.1663 & 0 \\ 0 & 0.0043 \end{bmatrix}$$

$$V = \begin{bmatrix} 0.7208 & -0.6932 \\ 0.6932 & 0.7208 \end{bmatrix}$$

Compared with either the threshold value or the relative error, it is clearly reasonable to assume that $rank(\underline{G}) = 1$, and no exact solution of the matrix $\underline{D}$ exists. Thus, the dimension of $\underline{G}$ must be increased by adding more Markov parameter matrices. Before collecting more Markov parameters, it is convenient for further analysis to modify $\hat{G}_0$ by setting its smallest singular value to zero, $i.e.$,

$$\hat{G}'_0 = U\Sigma'V$$

$$= \begin{bmatrix} 0.4464 & 0.8948 \\ 0.8948 & -0.4464 \end{bmatrix} \begin{bmatrix} 3.1663 & 0 \\ 0 & 0 \end{bmatrix} \begin{bmatrix} 0.7208 & -0.6932 \\ 0.6932 & 0.7208 \end{bmatrix}$$

$$= \begin{bmatrix} 1.0187 & 0.9797 \\ 2.0422 & 1.9640 \end{bmatrix}$$

This is a reasonable approximation when a singular value is significantly smaller than other singular values and the matrix has been tested and found to be rank deficient. In fact, it is recommended to test the rank of each Markov parameter matrix and modify them accordingly before one forms the block-Toeplitz matrix $\underline{G}$ and the block matrix Λ. Now increase the dimension of $\underline{G}$ by

$$\underline{G} = \begin{bmatrix} \hat{G}'_0 & 0 \\ \hat{G}_1 & \hat{G}'_0 \end{bmatrix} \tag{5.14}$$

The SVD decomposition yields

$$\Sigma = \begin{bmatrix} 4.4745 & 0 & 0 & 0 \\ 0 & 2.2556 & 0 & 0 \\ 0 & 0 & 0.0687 & 0 \\ 0 & 0 & 0 & 0 \end{bmatrix}$$

$$V = \begin{bmatrix} 0.6074 & -0.3866 & 0.6940 & 0 \\ 0.5474 & -0.4294 & -0.7183 & 0 \\ -0.4150 & -0.5883 & 0.0355 & -0.6932 \\ -0.3991 & -0.5657 & 0.0341 & 0.7208 \end{bmatrix}$$

Clearly, $rank(\underline{G}) \geq 2$ and

$$V_{21} = \begin{bmatrix} 0 \\ 0 \end{bmatrix}$$

An exact solution of $\underline{D}$ exists and the order of the interactor matrix $d = 2$ is selected.

5.6 Simulation and experimental evaluation on a pilot scale process

The block matrix is formed as

$$\hat{\Lambda} = \begin{bmatrix} 1.0187 & 2.0422 & -0.9321 & -1.4394 \\ 0.9797 & 1.9604 & -0.8885 & -1.1543 \end{bmatrix}^T$$

Following the algorithm in (Rogozinski, Paplinski, and Gibbard 1987) (and see Appendix A), the estimated unitary interactor matrix can be factored as

$$\hat{D} = \begin{bmatrix} -0.4464q & -0.8948q \\ -0.8948q^2 & 0.4464q^2 \end{bmatrix}$$

This result agrees well with the theoretical open-loop interactor matrix shown in Equation 5.12.

Example 5.6.2. To evaluate the proposed algorithm on a physical process, both open-loop and closed-loop experiments have been conducted on a two-interacting-tank pilot-scale process. Each tank is a double-walled glass tank, 50 cm high with an inside diameter of 14.5 cm. The levels (h_1, h_2) of the two tanks are the two controlled variables. The signals to the two valves (u_1, u_2) are manipulated to control the levels. The process is shown in Figure 5.2. The sampling interval is taken as $T_s = 20$s. A two-step delay (including zero-order-hold) is introduced at the first input, and a three-step delay (including zero-order-hold) is introduced at the second input. Two sets of multiloop P/PI controllers are implemented in the experiments. Open-loop and closed-loop interactor matrices are estimated from open-loop and closed-loop data respectively. The estimated open-loop and closed-loop interactor matrices are then compared.

An open-loop multivariable test was conducted with the result shown in Figure 5.3. The two manipulated signals, u_1 and u_2, are applied simultaneously. The multivariate prediction error method (Ljung 1998) is used for the identification of this multivariate process, which yields the following Markov parameters:

$$T_{open} = \begin{bmatrix} 0.1632 & 0 \\ 0.0169 & 0 \end{bmatrix} q^{-2} + \begin{bmatrix} 0.1462 & 0.0185 \\ 0.0355 & 0.1249 \end{bmatrix} q^{-3} + \cdots$$

Following the procedure as introduced in the foregoing sections, a unitary interactor matrix can be calculated as

$$D_{open} = \begin{bmatrix} -0.9947q^2 & -0.1030q^2 \\ -0.1030q^3 & 0.9947q^3 \end{bmatrix} \quad (5.15)$$

The interactor matrix depends on the first few steps of the initial responses. Since the initial responses of the interaction terms between the two tanks are relatively weak, the interactor matrix is dominated by the diagonal terms.

Closed-loop tests with a P-controller (Proportional only) and with a PI-controller are conducted respectively, yielding the results shown in Figure 5.4.

Dither signals, w_1 and w_2, are applied to the process simultaneously. The following Markov parameters of the closed-loop process are obtained from the closed-loop data:

$$T_P = \begin{bmatrix} 0.1910 & 0 \\ 0.0192 & 0 \end{bmatrix} q^{-2} + \begin{bmatrix} 0.1600 & 0.0090 \\ 0.0414 & 0.1260 \end{bmatrix} q^{-3} + \cdots$$

and

$$T_{PI} = \begin{bmatrix} 0.1704 & 0 \\ 0.0177 & 0 \end{bmatrix} q^{-2} + \begin{bmatrix} 0.1451 & 0.0132 \\ 0.0369 & 0.1290 \end{bmatrix} q^{-3} + \cdots$$

These yield the closed-loop interactor matrices as

$$D_P = \begin{bmatrix} -0.9950q^2 & -0.1000q^2 \\ -0.1000q^3 & 0.9950q^3 \end{bmatrix} \tag{5.16}$$

and

$$D_{PI} = \begin{bmatrix} -0.9946q^2 & -0.1033q^2 \\ -0.1033q^3 & 0.9946q^3 \end{bmatrix} \tag{5.17}$$

The similarity between Equation 5.15 and Equations 5.16 and 5.17 clearly demonstrates that the interactor matrix is "feedback-invariant" and can be estimated from closed-loop data. The small differences between these three interactor matrices may be attributed to disturbances.

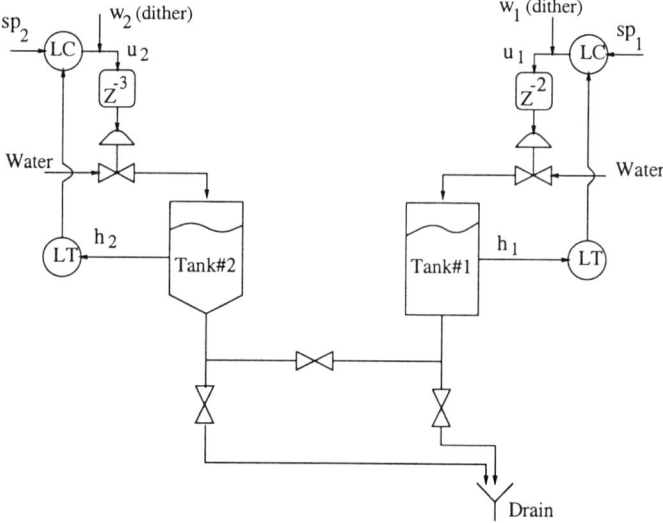

Fig. 5.2. Schematic of the two-interacting-tank pilot-scale process.

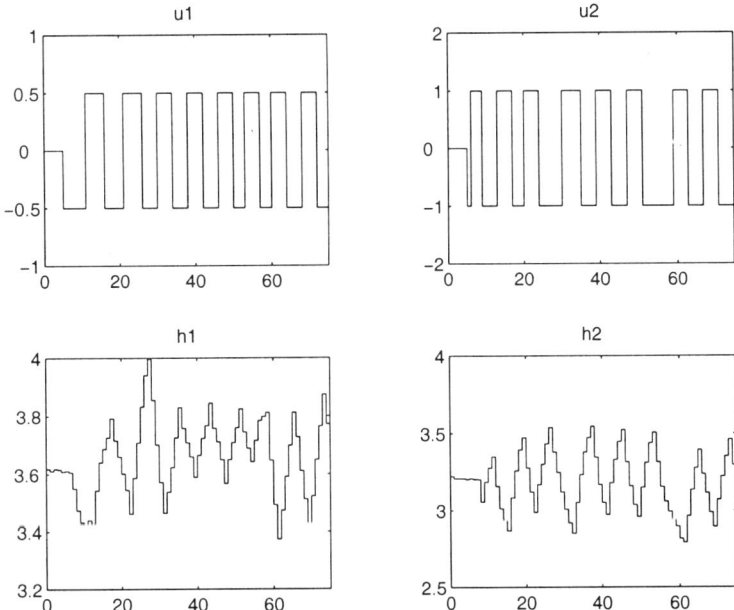

Fig. 5.3. Open-loop (input and output) test data where $u = 0$ corresponds to 50% opening of the valve, and the units of h_1 and h_2 are voltage. The time scale is in terms of sampling intervals.

5.7 Industrial application

Example 5.7.1. A multivariate industrial distillation process is studied in this example. A unitary interactor matrix is estimated from industrial closed-loop data.

The process consists of 3 distillation towers as shown in Figure 5.5. The bottom product of the last tower is the main product, and the distillates or the top products of the remaining two towers are recycled to the upstream process. The problem encountered for the multiloop controller design of this process is the strong interaction between levels of the first two towers and between the temperature and the level of the second tower.

Since the last tower is relatively small, temperature variation in the second tower can disturb the temperature of the last tower significantly. Maintaining the temperature of the last tower at a constant value is important for regulating the quality of the final product. Owing to the strong interaction between the temperature and the level loops in the second tower, a multivariable controller is clearly desirable. Therefore, the control objective in this study is the temperature and level control of the second tower in order to reduce disturbances to the last tower. To design a high performance

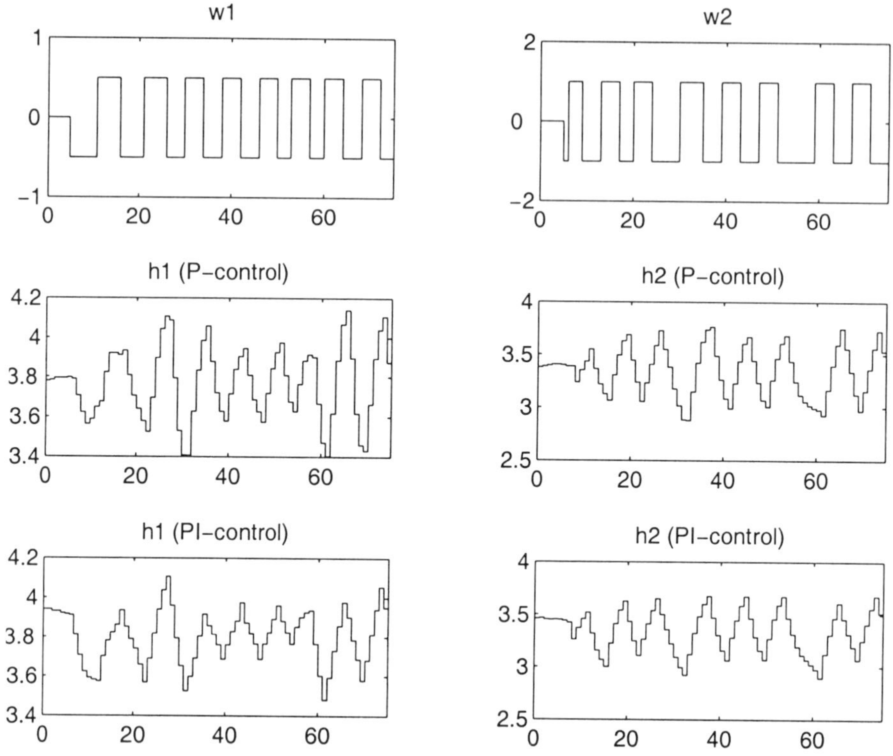

52   5. Estimation of the Unitary Interactor Matrices

Fig. 5.4. Closed-loop (dither and output) test data where the units of h_1 and h_2 are voltage. The time scale is in terms of sampling intervals.

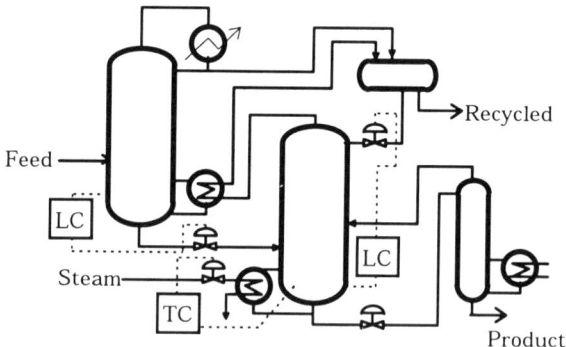

Fig. 5.5. The industrial process flowsheet

multivariable controller, it is in the interest of control engineers to know the interactor matrix.

Simple closed-loop setpoint changes were conducted on this process. To simplify the test, setpoint changes of the level and temperature were conducted separately. Identification of the closed-loop Markov parameters or the transfer function matrix can therefore be cast as four separate open-loop identification problems of SISO impulse response parameters or transfer functions. Figure 5.7 shows the four SISO identification results. The corresponding setpoints are shown in Figure 5.6. To identify closed-loop transfer function matrices of the temperature and level, only two experiments are required. One is for the level setpoint test, and the other one is for the temperature setpoint test. In this example, however, the level setpoint tests were conducted twice with different excitation signals owing to saturated data record of the temperature response when the first set of level setpoint tests was conducted. Note that in Figure 5.7, the solid line denotes the estimated-model prediction based only on the past inputs, and the dash-doted line denotes actual outputs.

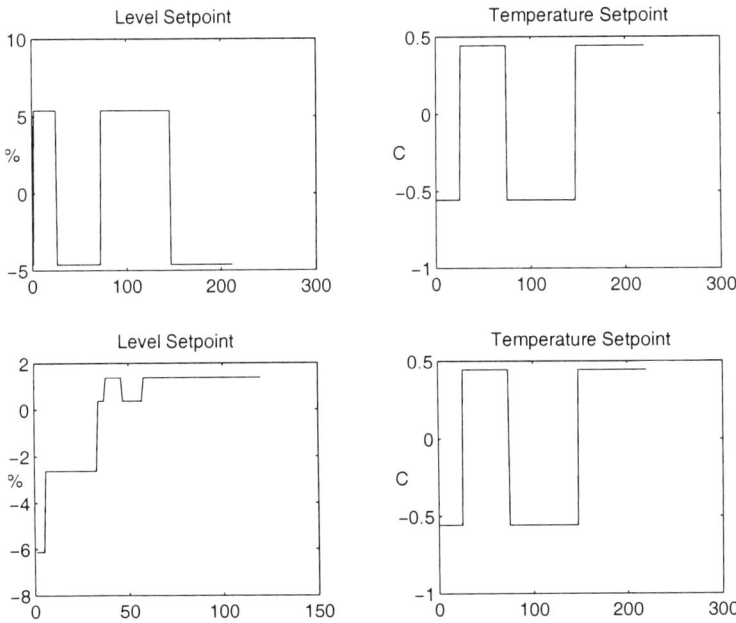

Fig. 5.6. Setpoints for the closed-loop tests. The time scale is in terms of sampling intervals

The four identified SISO models form the closed-loop transfer function matrix and yield the following Markov parameters.

5. Estimation of the Unitary Interactor Matrices

$$\hat{G} = \begin{bmatrix} 0 & 0 \\ 0 & 0.1513 \end{bmatrix} q^{-1} + \begin{bmatrix} 0 & -0.6629 \\ 0.0130 & 0.1267 \end{bmatrix} q^{-2} +$$

$$+ \begin{bmatrix} 0.0713 & -0.5022 \\ 0.0118 & 0.1062 \end{bmatrix} q^{-3} + \cdots$$

Since time-delays of each SISO model can be easily determined through SISO identification procedure (Soderstrom and Stoica 1989), zeros which appear in the above Markov parameter matrices are exact zeros and correspond to time-delays in the SISO models.

Following the same procedure as introduced earlier, an order of the interactor matrix $d = 3$ is obtained. The block matrix of the first three Markov parameters is formed as

$$\hat{\Lambda} = \begin{bmatrix} 0 & 0 & 0 & 0.0130 & 0.0713 & 0.0118 \\ 0 & 0.1513 & -0.6629 & 0.1267 & -0.5022 & 0.1062 \end{bmatrix}^T$$

The unitary interactor matrix is calculated as

$$D = \begin{bmatrix} 0.9749q^2 & -0.2225q \\ 0.2225q^3 & 0.9749q^2 \end{bmatrix}$$

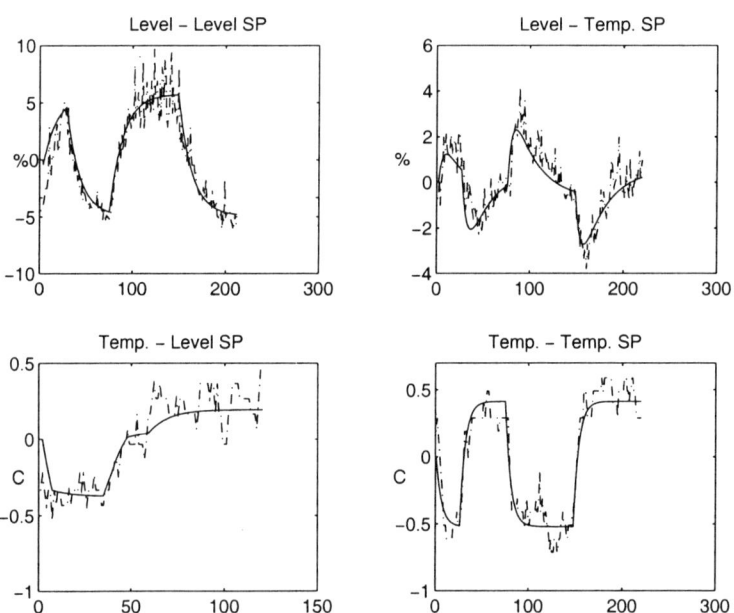

Fig. 5.7. Predicted and actual outputs; all data have been zero-mean centered. The time scale is in terms of sampling intervals

5.8 Summary

In this chapter, an algorithm has been developed for estimating interactor matrices. It is shown that the unitary interactor matrix is invariant under open or closed-loop conditions. The singular value decomposition method has been used to determine the order of the interactor matrix. The algorithm for the factorization of the unitary interactor matrix has been simplified by using only the first few Markov parameter matrices. The proposed algorithm has been evaluated by simulated example, pilot-scale experiment and industrial processes. The results presented in this chapter are useful for the design of minimum variance or singular LQ control, optimal H_2-control, optimal filtering, and particularly, multivariable control loop performance assessment methods.

CHAPTER 6
FEEDBACK CONTROLLER PERFORMANCE ASSESSMENT: SIMPLE INTERACTOR

6.1 Introduction

The interactor matrix D can be one of the three forms discussed in Chapter 3. If D is of the form: $D = q^d I$, then the transfer function matrix T is regarded as having a *simple interactor matrix*. This is the simplest form of the interactor matrices. Although it is usually unlikely to encounter a real process with a simple interactor matrix, the result presented in this chapter provides a basis for solutions to processes with diagonal and general interactor matrices that follow later.

This chapter is organized as follows: The feedback controller-invariance property of the minimum variance control term is discussed in Section 6.2. The FCOR algorithm, for the multivariate case, is presented in Section 6.3. The proposed algorithm is illustrated by a simulated example in Section 6.4, followed by concluding remarks in Section 6.5

6.2 Feedback controller-invariance of minimum variance term

The simplest form of a multivariable process has a square process transfer function matrix with a simple interactor matrix. Keviczky (Keviczky and Hetthessy 1977) and Borison (Borison 1979) have given the minimum variance control law for processes with simple interactor matrices. The purpose of this section is to show that the minimum variance term is feedback control invariant and can be estimated from routine operating data.

Theorem 6.2.1. *For the multivariable process with a simple interactor matrix*

$$Y_t = TU_t + Na_t \qquad (6.1)$$

the minimum variance control is obtained by minimizing

$$J = E[Y_t^T Y_t]$$

The performance measure is determined by the following steps:

1. The quadratic measure of minimum variance is determined by
$$E[Y_t^T Y_t]_{min} = E(e_t^T e_t) = tr(Var(Fa_t))$$
and the minimum variance itself is determined by
$$Var(Y_t)|_{min} = Var(e_t) = Var(Fa_t)$$
where $e_t = Fa_t$. The polynomial matrix F depends only on the time-delay d and the disturbance model, and satisfies the following identity:
$$q^{-d}DN = \underbrace{F_0 + \cdots + F_{d-1}q^{-(d-1)}}_{F} + q^{-d}R$$
where R is a proper rational transfer function matrix.

2. Furthermore, if one models closed-loop routine operating data under feedback control by the following multivariate moving-average process:
$$Y_t - E(Y_t) = \underbrace{F_0 a_t + F_1 a_{t-1} + \cdots + F_{d-1} a_{t-d+1}}_{e_t}$$
$$+ \underbrace{L_0 a_{t-d} + L_1 a_{t-d-1} + \cdots}_{w_{t-d}}$$
then the minimum variance term, e_t, consists of the first d terms of this moving-average model, and, therefore can be separated out by using time series analysis of routine operating data and used as a benchmark measure of multivariate minimum variance control.

Proof. For this case of the simple interactor matrix, the transfer function matrix can be written as
$$T = q^{-d}\tilde{T} \quad (6.2)$$
where $\tilde{T}$ is the delay-free transfer function matrix of T. Substituting Equation 6.2 into 6.1 yields
$$Y_t = q^{-d}\tilde{T}U_t + Na_t$$
Consider the feedback control law given by $U_t = -QY_t$. The closed-loop transfer function is then determined by
$$Y_t = -q^{-d}\tilde{T}QY_t + Na_t \quad (6.3)$$
Now consider the Diophantine identity:
$$N = F + q^{-d}R \quad (6.4)$$
where
$$F = F_0 + F_1 q^{-1} + \ldots + F_{d-1} q^{-(d-1)}$$
and R is the remaining proper and rational transfer function matrix. Substituting Equation 6.4 into 6.3 gives

6.2 Feedback controller-invariance of minimum variance term

$$Y_t = (q^d I + \tilde{T}Q)^{-1} q^d (F + q^{-d} R) a_t$$

Applying the matrix inverse lemma yields

$$\begin{aligned}
Y_t &= [q^{-d}I - q^{-d}\tilde{T}(I + q^{-d}Q\tilde{T})^{-1} Q q^{-d}] q^d [F + q^{-d} R] a_t \\
&= F a_t - q^{-d}\tilde{T}(I + q^{-d}Q\tilde{T})^{-1} Q F a_t + q^{-d} R a_t - \\
&\quad - q^{-2d}\tilde{T}(I + q^{-d}Q\tilde{T})^{-1} Q R a_t \\
&= F a_t + q^{-d} R a_t - q^{-d}\tilde{T}(I + q^{-d}Q\tilde{T})^{-1} Q N a_t \\
&\triangleq F a_t + L a_{t-d}
\end{aligned}$$ (6.5)

where

$$L = R - \tilde{T}(I + q^{-d}Q\tilde{T})^{-1} Q N \tag{6.6}$$

is a proper rational transfer function matrix. The two terms on the right hand side of Equation 6.5 are therefore independent.

Now define $e_t = F a_t$ and $w_{t-d} = L a_{t-d}$. Then

$$Var(Y_t) \geq Var(e_t) = Var(F a_t)$$

and

$$E[Y_t^T Y_t] \geq E[e_t^T e_t] = tr(Var(F a_t))$$

The equality holds under minimum variance control when $L = 0$. The minimum variance control law is therefore obtained by simply setting $L = 0$ in Equation 6.6. The resulting controller transfer function, $U_t = -QY_t$ is determined by

$$Q = -\tilde{T}^{-1}(q^{-d}I - NR^{-1})^{-1}$$

Combining this with Equation 6.4 yields

$$\begin{aligned}
Q &= -\tilde{T}^{-1}[q^{-d}I - (F + q^{-d}R)R^{-1}]^{-1} \\
&= \tilde{T}^{-1} R F^{-1}
\end{aligned}$$ (6.7)

which is the minimum variance control law. Notice that in Equation 6.5, the controller has no influence on the first term, which is the minimum variance term of the process output and is given by

$$Y_t|_{mv} = e_t = F a_t = (F_0 + F_1 q^{-1} + \ldots + F_{d-1} q^{-(d-1)}) a_t \tag{6.8}$$

Therefore, if a closed-loop response under feedback control is modelled by a multivariate moving-average process as

$$\tilde{Y}_t - E(\tilde{Y}_t) = \underbrace{F_0 a_t + F_1 a_{t-1} + \cdots + F_{d-1} a_{t-d+1}}_{e_t} + \underbrace{L_0 a_{t-d} + L_1 a_{t-d-1} + \cdots}_{w_{t-d}}$$ (6.9)

then, the feedback control invariant term, e_t, can be separated out from other terms that can be influenced by feedback control. The performance of minimum variance control is subsequently calculated from e_t.

6.3 The FCOR algorithm

6.3.1 Multivariable performance index

As proved in the last section, $e_t \triangleq Fa_t$ is the feedback control invariant minimum variance term. This minimum variance term can be used as a benchmark for the multivariable performance measure. By using Equation 6.8, one can write the minimum variance term e_t as

$$e_t = (F_0 + q^{-1}F_1 + \cdots + q^{-d+1}F_{d-1})a_t$$

Thus

$$\begin{aligned} \Sigma_{mv} &= E[e_t e_t^T] \\ &= F_0 \Sigma_a F_0^T + \cdots + F_{d-1} \Sigma_a F_{d-1}^T \end{aligned} \quad (6.10)$$

where

$$\Sigma_a = E(a_t a_t^T)$$

On the other hand, the closed-loop output vector Y_t under feedback control can be represented by an infinite multivariate moving average process, i.e.,

$$Y_t = F_0 a_t + F_1 a_{t-1} + \cdots + F_{d-1} a_{t-d+1} + F_d a_{t-d} + \cdots \quad (6.11)$$

so the covariance of the output and the noise at lag i is given by

$$\Sigma_{Ya}(i) = E[Y_t a_{t-i}^T] = F_i \Sigma_a \quad (6.12)$$

Now consider the definition of the performance index as

$$\eta(d) \triangleq tr(\Sigma_{mv} \tilde{\Sigma}_Y^{-1})/n \quad (6.13)$$

where n is the dimension of Y_t, $\tilde{\Sigma}_Y = diag(\Sigma_Y)$ and $\Sigma_Y = Var(Y_t)$. When a process is under minimum variance control, we have $\Sigma_Y = \Sigma_{mv}$; thus, it can be shown that $\eta(d) = 1$. If control is poor relative to the minimum variance control, then we should expect $0 < \eta(d) < 1$. Applying Equation 6.13 to a SISO process where $n = 1$ gives

$$\eta(d) = tr(\frac{\sigma_{mv}^2}{\sigma_y^2}) = \frac{\sigma_{mv}^2}{\sigma_y^2}$$

which is the performance measure of a SISO process. A correlation analysis yields a computationally simple procedure for calculating $\eta(d)$ as is discussed next.

Equation 6.13 can be written as

$$\begin{aligned} n \times \eta(d) &= tr(\Sigma_{mv} \tilde{\Sigma}_Y^{-1}) \\ &= tr(\tilde{\Sigma}_Y^{-1/2} \Sigma_{mv} \tilde{\Sigma}_Y^{-1/2}) \end{aligned} \quad (6.14)$$

6.3 The FCOR algorithm

Substituting Equations 6.10 into 6.14 and using the relation established in Equation 6.12 yields

$$
\begin{aligned}
n \times \eta(d) &= tr[\tilde{\Sigma}_Y^{-1/2} \Sigma_{Ya}(0) \Sigma_a^{-1} \Sigma_{aY}(0) \tilde{\Sigma}_Y^{-1/2} + \\
&\quad + \tilde{\Sigma}_Y^{-1/2} \Sigma_{Ya}(1) \Sigma_a^{-1} \Sigma_{aY}(1) \tilde{\Sigma}_Y^{-1/2} + \\
&\quad + \cdots + \tilde{\Sigma}_Y^{-1/2} \Sigma_{Ya}(d-1) \Sigma_a^{-1} \Sigma_{aY}(d-1) \tilde{\Sigma}_Y^{-1/2}] \\
&= tr(\rho_{Ya}(0) \rho_a^{-1} \rho_{aY}(0) + \rho_{Ya}(1) \rho_a^{-1} \rho_{aY}(1) + \cdots + \\
&\quad + \rho_{Ya}(d-1) \rho_a^{-1} \rho_{aY}(d-1))
\end{aligned}
$$

where $\rho_{Ya}(i) = \tilde{\Sigma}_Y^{-1/2} \Sigma_{Ya}(i) \tilde{\Sigma}_a^{-1/2}$, ($\tilde{\Sigma}_a \triangleq diag(\Sigma_a)$), is the multivariate cross correlation between Y_t and a_{t-i}; and $\rho_a = \tilde{\Sigma}_a^{-1/2} \Sigma_a \tilde{\Sigma}_a^{-1/2}$ is the multivariate autocorrelation of a_t. If a scaled cross correlation is defined as $\bar{\rho}_{Ya}(i) \triangleq \rho_{Ya}(i) \rho_a^{-1/2}$ and

$$ Z \triangleq [\ \bar{\rho}_{Ya}(0) \quad \bar{\rho}_{Ya}(1) \quad \cdots \quad \bar{\rho}_{Ya}(d-1)\ ] $$

then we have

$$ \eta(d) = tr Z Z^T / n \tag{6.15} $$

Note from Equations 6.14 and 6.15 that

$$ ZZ^T = \tilde{\Sigma}_Y^{-1/2} \Sigma_{mv} \tilde{\Sigma}_Y^{-1/2} $$

The above multivariable performance thus has a clear physical interpretation which is stated as follows: The diagonal elements of matrix ZZ^T are the performance measures of each single output, $\eta_{Y1}(d), \cdots, \eta_{Yn}(d)$, and therefore

$$
\begin{aligned}
\eta(d) &= tr(ZZ^T)/n = \frac{\eta_{Y1}(d) + \cdots + \eta_{Yn}(d)}{n} \\
&= \text{(the average performance of n outputs)}
\end{aligned}
$$

where the individual performance, $\eta_{Yi}(d)$, is the ratio of the minimum variance (under multivariable minimum variance control) and the actual variance in the ith output. The performance measure is therefore a comparison between the variance of each single output and that of the corresponding minimum variance output achieved under multivariable minimum variance control. Thus, one can obtain the single output performance index from the diagonal elements of the matrix ZZ^T simultaneously when we calculate the overall multivariable performance index $\eta(d)$. Furthermore, if we take the process offset into account, the modified performance index can be written as

$$ \eta'(d) = tr(\Sigma_{mv} \tilde{\Sigma}_{mse}^{-1})/n \tag{6.16} $$

where $\tilde{\Sigma}_{mse}$ is the diagonal matrix of Σ_{mse}, and $\Sigma_{mse} = \Sigma_Y + \delta\delta^T$ is the mean square error. The expression for $\eta'(d)$ can also be simplified by following the analogous procedure for $\eta(d)$ that results in Expression 6.15 from Equation 6.14.

Although a_t is unknown, it can be replaced by the estimated white noise $\hat{a}_t$, using a similar filtering/whitening procedure as introduced in Chapter 2, but by replacing the scalar output y_t with a vector output Y_t in the filtering/whitening process, the univariate time series analysis has also to be replaced by multivariate time series analysis. The FCOR algorithm provides a relatively easy way to calculate the performance measure of a multivariable process with a simple interactor matrix, but the algorithm is not limited to the simple interactor matrix process. In the following chapters, it will be shown that, via interactor filtering of the outputs, multivariable processes with diagonal or triangular interactor matrices can be transformed to the simple interactor matrix form. This is convenient for computing the minimum variance control law and the performance indices.

6.4 Simulation

Example 6.4.1. This example illustrates the performance assessment of a multivariable process with a simple interactor matrix.

Consider a multivariable process: $Y_t = TU_t + Na_t$, where

$$T = q^{-2} \begin{bmatrix} \frac{1}{1-0.4q^{-1}} & \frac{2}{1-0.5q^{-1}} \\ \frac{1}{1-0.1q^{-1}} & \frac{1}{1-0.2q^{-1}} \end{bmatrix}$$

$$N = \begin{bmatrix} \frac{2}{1-0.9q^{-1}} & \frac{1}{1-0.3q^{-1}} \\ \frac{1}{1-0.4q^{-1}} & \frac{2}{1-0.5q^{-1}} \end{bmatrix}$$

Because

$$\lim_{q^{-1} \to 0} q^2 T = \begin{bmatrix} 1 & 2 \\ 1 & 1 \end{bmatrix}$$

is of full rank, we conclude that $d = 2$. Using Equation 6.4, N is separated to

$$N = \begin{bmatrix} 2 + 1.8q^{-1} & 1 + 0.3q^{-1} \\ 1 + 0.4q^{-1} & 2 + q^{-1} \end{bmatrix} + \begin{bmatrix} \frac{1.62q^{-2}}{1-0.9q^{-1}} & \frac{0.09q^{-2}}{1-0.3q^{-1}} \\ \frac{0.16q^{-2}}{1-0.4q^{-1}} & \frac{0.5q^{-2}}{1-0.5q^{-1}} \end{bmatrix}$$

$$\stackrel{\triangle}{=} F + q^{-2}R \tag{6.17}$$

Under minimum variance control, the process output is written as $e_t = Fa_t$, and this part of the output is invariant under feedback control and can be used as a benchmark to access controller performance for this process.

For the sake of illustration, consider a simple proportional control of the form

$$U_t = -QY_t = -\begin{bmatrix} K & 0 \\ 0 & K \end{bmatrix} Y_t$$

Applying the FCOR algorithm to the output vector Y_t yields the results shown in Figure 6.1. The solid line and the dashed line denote the theoretical value of the performance indices with and without considering the offset respectively while the asterisk and circle denote the corresponding estimated performance indices. The abscissa in this graph corresponds to the value of the proportional gain. It can be seen that for this form of proportional, diagonal control, the maximum performance measure is 0.7 when the proportional gain K takes the value of 0.1 if the offset is not considered in the performance measure. The performance measure under open-loop condition (when $K = 0$) appears to be acceptable compared with the maximum performance measure (0.7). If the offset of a process output is also considered in

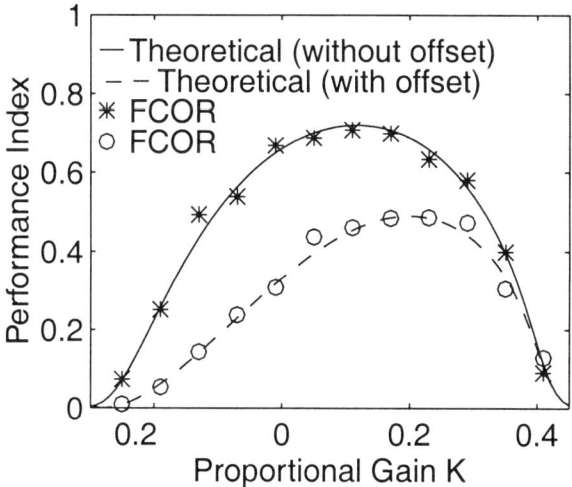

Fig. 6.1. Simple interactor matrix MIMO performance assessment. Each asterisk represents an estimation based on 1000 data points using the FCOR algorithm.

the measure of performance, a deteriorated performance would be expected. As shown by the dashed line (the setpoint is taken as $Y_t^{sp} = [3,3]^T$ in the simulation), the overall performance index is now lower than the performance measure without considering the offset. The open loop process (when $K = 0$) no longer demonstrates good performance. The best performance is obtained when $K = 0.2$ instead of $K = 0.1$. Increasing the proportional gain decreases the offset and, therefore, improves the performance when the offset dominates the output. This is the reason that the best performance shifts toward the right from $K = 0.1$ to $K = 0.2$, but with a gain greater than 0.2, the process becomes more oscillatory and, eventually, unstable. When the variation caused by the oscillation dominates the output, the performance goes down rapidly. The offset no longer dominates the performance measure and this measure now approaches the same measure without considering the offset.

6.5 Summary

The multivariable performance measure has been defined, and the computationally simple algorithm (the FCOR algorithm) that estimates multivariate performance indices has been established in this chapter. The results can be applied to multivariable processes with simple interactor matrices, and they provide a basis for the following chapters. The application of the proposed algorithm has been demonstrated by a simulated example.

CHAPTER 7
FEEDBACK CONTROLLER PERFORMANCE ASSESSMENT: DIAGONAL INTERACTOR

7.1 Introduction

Although minimum variance control is not practically desirable owing to its poor robustness and/or excessive control effort requirement, it does provide an absolute lower bound on the process variance. This lower bound naturally serves as a useful benchmark to evaluate current control loop performance if reduction of process variation is the control objective. Such a control loop performance measure provides guidelines and useful information for control engineers when they design, tune or upgrade controllers or control strategies. If the best performance cannot satisfy the requirement, alternative control strategies such as implementing feedforward control and/or reducing dead time may be necessary. For a number of industrial processes (particularly pulp/paper processes), reduction of process variation is the main objective in controller design. Performance assessment with minimum variance control as the benchmark is, therefore, particularly useful for such processes. In fact, the first application of the performance assessment technique was on a paper machine (Astrom 1967), but most industrial processes are inherently multivariate in nature so performance assessment with multivariate minimum variance control as the benchmark is more desirable. This chapter is an extension of Chapter 6 by considering closed-loop performance assessment of multivariate processes with the diagonal interactor matrix.

In this chapter, feedback invariance of multivariable processes with diagonal interactor matrix is discussed in Section 7.2.1, followed by a brief extension of the FCOR algorithm in Section 7.3. The chapter also considers a detailed evaluation of an industrial headbox control system using routine operating data in Sections 7.4 and 7.5.

7.2 Feedback controller-invariance of minimum variance term

7.2.1 Feedback controller-invariance of minimum variance term

Chapter 2 has shown that the measure of minimum variance control performance of SISO processes can be estimated from routine operating data. The key to this property is that the minimum variance term is feedback control invariant. This idea has been extended to MIMO processes with the simple interactor matrix in Chapter 6.

For MIMO processes with a general interactor matrix (i.e., neither simple nor diagonal), the feedback invariance property of minimum variance control (minimizing $J = E(Y_t - Y_t^{sp})^T(Y_t - Y_t^{sp})$ or $J = E(Y_t - Y_t^{sp})^T W (Y_t - Y_t^{sp})$) can also be solved by using the unitary or weighted unitary interactor matrix as will be discussed in Chapter 8. Just as *a priori* knowledge of the time delay is required for SISO applications, performance assessment of MIMO processes with general interactor matrices requires that the interactor matrix must be known from *a priori* knowledge. This is tantamount to knowing the entire transfer function matrix (Goodwin and Sin 1984) or, at least, the first few Markov parameters or impulse response coefficients of the transfer function matrix as discussed in Chapter 5.

However, some well-designed multivariable processes have the structure of the diagonal interactor. Like the SISO case, the diagonal interactor matrix only depends on the pure time delays of the transfer function and is easier to obtain from *a priori* knowledge of process; therefore, this diagonal structure is elaborated in the present chapter. The treatment of the general interactor matrix is discussed in Chapter 8.

In Chapter 6, we have shown that for the process with a simple interactor matrix

$$Y_t = q^{-d}\tilde{T}U_t + Na_t \tag{7.1}$$

where d is the time delay and $\tilde{T}$ is the delay-free transfer function matrix, the following inequality holds:

$$Var(Y_t) \geq E(e_t)(e_t)^T = Var(Fa_t)$$

where $e_t = Fa_t$, F is defined by the identity

$$N = \underbrace{F_0 + F_1 q^{-1} + \cdots + F_{d-1}q^{-(d-1)}}_{F} + q^{-d}R$$

F_i (for $i = 0, \cdots, d-1$) are also constant coefficient matrices, and R is a proper rational transfer function matrix. If closed-loop, routine operating data under feedback control is modelled by a multivariate moving-average process

$$Y_t - E(Y_t) = \underbrace{F_0 a_t + F_1 a_{t-1} + \cdots + F_{d-1}a_{t-d+1}}_{e_t} + \underbrace{L_0 a_{t-d} + L_1 a_{t-d-1} + \cdots}_{w_{t-d}}$$
(7.2)

where L_i (for $i = 0, 1, \cdots$) are constant coefficient matrices, then the term w_{t-d} is feedback control dependent, and the term, e_t, consisting of the first d terms of the moving-average model, is independent of feedback control. Under minimum variance control, w_{t-d} vanishes, and therefore e_t represents the process under minimum variance control and can be estimated from routine operating data.

Now consider a process with a diagonal interactor matrix,

$$Y_t = TU_t + Na_t = D^{-1}\tilde{T}U_t + Na_t \tag{7.3}$$

where

$$D = diag(q^{d_1}, q^{d_2}, \cdots, q^{d_n})$$

Multiplying both sides of Equation 7.3 by $q^{-d}D$, where $d = max(d_1, \cdots, d_n)$, yields

$$\begin{aligned} q^{-d}DY_t &= q^{-d}\tilde{T}U_t + q^{-d}DNa_t \\ &= q^{-d}\tilde{T}U_t + \tilde{N}a_t \end{aligned} \tag{7.4}$$

where $\tilde{N}$ is a proper transfer function matrix between the disturbance a_t and the interactor-filtered output $q^{-d}DY_t$. By defining $\tilde{Y}_t = q^{-d}DY_t$, Equation 7.4 has been transferred to the same form as Equation 7.1, i.e.,

$$\tilde{Y}_t = q^{-d}\tilde{T}U_t + \tilde{N}a_t \tag{7.5}$$

This is a process with a simple interactor matrix. It follows that

$$Var(\tilde{Y}_t) \geq E(\tilde{e}_t)(\tilde{e}_t)^T = Var(\tilde{F}a_t)$$

where $\tilde{e}_t = \tilde{F}a_t$, and $\tilde{F}$ is defined by the identity:

$$\tilde{N} = \underbrace{\tilde{F}_0 + \tilde{F}_1 q^{-1} + \cdots + \tilde{F}_{d-1}q^{-(d-1)}}_{\tilde{F}} + q^{-d}\tilde{R}$$

Thus, if the interactor-filtered, routine operating data under feedback closed-loop control is modelled by a multivariate moving-average process:

$$\tilde{Y}_t - E(\tilde{Y}_t) = \underbrace{\tilde{F}_0 a_t + \tilde{F}_1 a_{t-1} + \cdots + \tilde{F}_{d-1}a_{t-d+1}}_{\tilde{e}_t} + \underbrace{\tilde{L}_0 a_{t-d} + \tilde{L}_1 a_{t-d-1} + \cdots}_{\tilde{w}_{t-d}}$$

(7.6)

then the minimum variance term $\tilde{e}_t$ is independent of feedback control and can be estimated from routine operating data, therefore.

Although the feedback invariance term $\tilde{e}_t$ represents the minimum variance term of the interactor-filtered variable $\tilde{Y}_t$, it also represents the minimum variance term of the original variable Y_t. Dugard et al. (1984) have shown that for the case of the diagonal interactor matrix, the control law which minimizes variance of the interactor-filtered variable $\tilde{Y}_t$ also minimizes variance of each element of the original variable Y_t (i.e., $y_i(t)$, for $i = 1, \cdots, n$). Thus, the diagonal elements of $Var(\tilde{e}_t)$ also provide absolute lower bounds of variance for each original output under multivariable feedback control. Furthermore, note that $\tilde{y}_i = q^{-d+d_i}y_i$, and therefore

$$Var(y_1(t)) = Var(\tilde{y}_1(t))$$
$$\vdots$$
$$Var(y_n(t)) = Var(\tilde{y}_n(t))$$

i.e., the diagonal elements of the variance (covariance) matrix of the original variable Y_t are the same as that of the interactor-filtered variable $\tilde{Y}_t$. As shown in Chapter 6, the diagonal elements of the variance (covariance) matrix are the variance of each output and are the terms required in performance assessment. Thus, performance assessment of the original variable Y_t is equivalent to performance assessment of the interactor-filtered variable $\tilde{Y}_t$.

7.2.2 Effect of non-minimum phase zeros

One natural question is how one evaluates performance of processes that have non-minimum phase zeros. This question also applies to the previous chapter and the following chapter. Desborough and Harris (1992) and Huang et al. (1996b) have pointed out that this does not affect its application. Performance assessment simply provides an absolute lower bound of process variance, although the lower bound may or may not be practically realizable or admissible depending on the zero location of the process. This information concerning the absolute lower bound is particularly useful for the design, tuning, monitoring and upgrading of control loops. For loops which indicate high performance measures, further tuning of controllers is neither necessary nor useful. For loops which indicate poor performance measures, further analysis such as process identification and/or re-design of the control algorithm may be necessary. The existence of non-minimum phase zeros implies that the actual or achievable lower bound is larger than the absolute lower bound. Consequently, the performance measure underestimates the actual or achievable performance. Take the performance measure of a SISO process as an example. This performance measure or index is defined as $\eta(d) = \sigma_{mv}^2/\sigma_y^2$, where σ_{mv}^2 is the absolute lower bound of process variance, and σ_y^2 is the variance of the process output. If a process has non-minimum phase zeros, then its achievable minimum variance is $\tilde{\sigma}_{mv}^2$ with $\tilde{\sigma}_{mv}^2 > \sigma_{mv}^2$. Consequently,

$$\eta(d) = \sigma_{mv}^2/\sigma_y^2 < \tilde{\sigma}_{mv}^2/\sigma_y^2 = \eta'(d)$$

where $\eta'(d)$ is the achievable performance measure; therefore, the performance measure $\eta(d)$ underestimates the achievable performance $\eta'(d)$, but this does not affect its application. For example, if a process has an acceptable performance measure (e.g., $\eta(d) > 0.5$), design considerations owing to the existence of non-minimum phase zeros should, in fact, bolster its acceptability since its achievable performance measure is likely to be even higher than its absolute performance measure (i.e., $\eta'(d) > \eta(d) > 0.5$). On the other hand, if a process has an unacceptable performance measure, it falls into the category of processes which may require further analysis, e.g., identification or controller redesign; therefore, existence of non-minimum phase zeros does not matter since these non-minimum phase zeros can be detected via identification, and the achievable performance measure $\eta'(d)$ can be subsequently obtained anyway. Once again, it must be emphasized that the techniques proposed in this chapter and other literature, e.g., Harris (1989), only require routine operating data and *a priori* knowledge of time delays. The effect of non-minimum phase zeros will be discussed again in Chapter 10 in more details.

7.3 Performance measures

7.3.1 The FCOR algorithm

The multivariable performance index is defined in Chapter 6 as

$$\eta(d) \triangleq tr(\Sigma_{mv}\tilde{\Sigma}_Y^{-1})/n \qquad (7.7)$$

and performance indices of individual outputs or the individual performance indices are defined as

$$[\eta_{y_1}, \cdots, \eta_{y_n}]^T \triangleq diag(\Sigma_{mv}\tilde{\Sigma}_Y^{-1}) \qquad (7.8)$$

where $\tilde{\Sigma}_Y = diag(\Sigma_Y)$, $\Sigma_Y = Var(Y_t)$, and Σ_{mv} is the lower bound of Σ_Y. These indices indicate the comparison of variance between the diagonal elements of the actual variance matrix and the corresponding diagonal elements of the minimum variance matrix.

For the process with a diagonal interactor matrix, these performance indices are equivalent to performance indices of the interactor-filtered variable. Therefore, performance assessment of the original variable can be obtained from performance assessment of the interactor-filtered variable, *i.e.*,

$$\eta(d) = tr(\Sigma_{\bar{m}v}\tilde{\Sigma}_{\tilde{Y}}^{-1})/n \qquad (7.9)$$

and

$$[\eta_{y_1}, \cdots, \eta_{y_n}]^T = diag(\Sigma_{\bar{m}v}\tilde{\Sigma}_{\tilde{Y}}^{-1}) \qquad (7.10)$$

where $\tilde{\Sigma}_{\tilde{Y}} = diag(\Sigma_{\tilde{Y}})$, $\Sigma_{\tilde{Y}} = Var(\tilde{Y}_t)$, and $\Sigma_{\bar{m}v}$ is the lower bound of $\Sigma_{\tilde{Y}}$.

Chapter 6 has shown that multivariate correlation analysis yields a computationally simple procedure for calculating these performance indices as follows:

Define a scaled cross correlation as

$$\bar{\rho}_{\tilde{Y}a}(i) \triangleq \rho_{\tilde{Y}a}(i)\rho_a^{-1/2}$$

and a block matrix as

$$Z \triangleq [\bar{\rho}_{\tilde{Y}a}(0), \bar{\rho}_{\tilde{Y}a}(1), \cdots, \bar{\rho}_{\tilde{Y}a}(d-1)]$$

where $\rho_{\tilde{Y}a}(i) = \tilde{\Sigma}_{\tilde{Y}}^{-1/2} \Sigma_{\tilde{Y}a}(i) \tilde{\Sigma}_a^{-1/2}$ is the multivariate cross correlation between $\tilde{Y}_t$ and a_{t-i} (note $\tilde{\Sigma}_a = diag(\Sigma_a)$ and $\tilde{\Sigma}_{\tilde{Y}} = diag(\Sigma_{\tilde{Y}})$); and $\rho_a = \tilde{\Sigma}_a^{-1/2} \Sigma_a \tilde{\Sigma}_a^{-1/2}$ is the multivariate autocorrelation of a_t. Then

$$\eta(d) = tr(ZZ^T)/n \tag{7.11}$$

and

$$[\eta_{y_1}, \cdots, \eta_{y_n}]^T = diag\{ZZ^T\} \tag{7.12}$$

where $\eta(d)$ is in fact the average of the individual performance indices. Although a_t is unknown, it can be replaced by the estimated white noise $\hat{a}_t$ from a filtering process as shown in Chapter 6. This filtering procedure is equivalent to fitting Y_t by a multivariate time series (ARI or ARIMA). The residual after fitting is the "whitened" noise $\hat{a}_t$. This procedure, based on Filtering and Correlation analysis, is termed the FCOR algorithm. Thus, the FCOR approach provides a relatively easy way to calculate the performance measure of a multivariable process with a diagonal interactor matrix, which avoids solving the multivariate Diophantine identity, but the algorithm is not limited to processes with the diagonal interactor matrix. It will be extended to processes with the general interactor matrix in Chapter 8.

7.3.2 The effect of sampling intervals

To apply the FCOR algorithm, a representative set of routine operating data should be sampled. Theoretically, the data sampling frequency is assumed to be the same as the controller sampling frequency. This sampling frequency may not be always desirable in practice for the following reasons: 1) the quality measure of many industrial processes is based on the outputs sampled at other frequencies which can be higher or lower than the controller sampling frequency, and a different sampling frequency of a stochastic signal may result in a different measure of the variance (MacGregor 1976); 2) the controller sampling rate can be very fast or even continuous on some control loops such as PID loops, and minimum variance control using such fast controller sampling frequency usually requires an excessive control action and gives an unrealistic benchmark performance; 3) the different control loops may have different controller sampling frequencies, i.e., no unique sampling frequency may be available for multivariate performance assessment; 4) typically, many available industrial data are sampled at a lower frequency than the controller sampling frequency. In such cases, one has to use a data sampling frequency which is different from the controller sampling frequency. Since the feedback-invariance property holds for any causal linear feedback controller within the time-delay period, the FCOR algorithm is generally valid for any sampling frequency as well. It calculates a performance index relative to a minimum variance controller whose control sampling frequency is the same as the data sampling frequency, but a controller sampling frequency different from the data sampling frequency means that actual performance assessment should consider the different effects of an extra time delay in addition to the actual physical delay in the process. This additional time delay corresponds to the presence of a zero-order-hold device. To illustrate the point, we assume that the process time delay is t_d, the control sampling interval of the existing controller is t_c, and the data sampling interval or the control sampling interval of the assumed benchmark (minimum variance) control is t_s. Then the actual time-delay in the process is $t_d + t_c$, and the time-delay for the assumed benchmark (minimum variance) control is $t_d + t_s$. In order to separate the minimum variance portion from routine operating data, feedback invariance must hold within the time-delay period from 0 to $t_d + t_s$ in the existing control loops. Since feedback-invariance does hold within the actual time-delay period from 0 to $t_d + t_c$, the FCOR algorithm estimates a theoretically exact result if $t_d + t_s \leq t_d + t_c$ (if the data sampling frequency is higher than the controller sampling frequency). However, if $t_d + t_s > t_d + t_c$ (the data sampling frequency is lower than the controller sampling frequency), then the feedback-invariance property holds within the actual time-delay period from 0 to $t_d + t_c$, but does not hold from $t_d + t_c$ to $t_d + t_s$; therefore, $t_d + t_c$ (instead of $t_d + t_s$) is recommended as the approximate time-delay in the calculation of the performance index when the data sampling frequency is lower than the controller sampling frequency. This underestimates the time-delay by the portion $t_s - t_c$ and may underestimate the performance index consequently. As illustrated in the last section, however, slight underestimation of the performance index does not affect its application.

7.4 Application to a headbox control system

The headbox is the soul of the entire pulp/paper machine (Newcombe 1991). Its purpose is to transform a piped flow of pulp stock into a homogeneous, uniform flow across the width of a machine wire running at high speed. Weyerhauser's Grande Prairie NSK pulp mill utilizes a Fourdrinier machine commissioned in the early 1970's. The process description, control objectives and problem description are discussed next.

7.4.1 Process description

The headbox is a unit operation within the pulp/paper-making process which takes stock (pulp and water mixture) flowing in a pipe and transforms it into a uniform, rectangular flow, equal in width to the machine wire and uniform in velocity in the machine direction. Good headbox operation results in uniform basis weight, little or no flocculation, and excellent retention on the wire. The schematic of the Fourdrinier machine employed by the mill at Grande Prairie is shown in Figure 7.1. The pressure/vacuum-air-pad headbox is used to produce a sheet of approximately 149.7 kg/278.7 m² (330 lb/3000 ft²). White water mixed with thick stock is delivered to the bottom of the headbox by a high-speed fan pump. A rotating rectifier near the slice lip keeps the pulp evenly distributed across the machine. A vacuum pump is used to reduce the air pad pressure below atmospheric pressure.

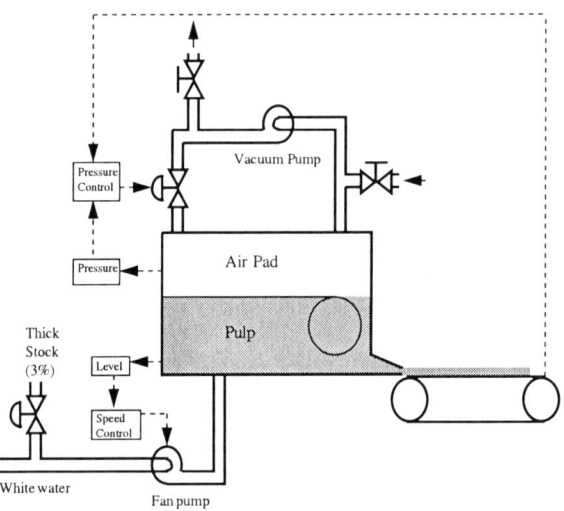

Fig. 7.1. Schematic diagram of the headbox

7.4.2 Process control

The primary control objective in the headbox control is to obtain a uniform basis weight, moisture, and caliper on the sheet. These properties are important for both the operability of the machine and the pulp's final quality. In keeping with common practice (Nordstrom and Norman 1994), sheet formation is maintained by continuous jet : wire ratio, headbox consistency and pond level control. The plant experiences poor operability when these control parameters deviate from their optimum values. Headbox consistency is controlled through thick stock (basis mass) valve adjustments, although the slice opening will impact consistency as well (Rice 1972). Pond level is controlled by changing the fan pump speed. The jet : wire ratio is a function of total head and wire speed. The total head at the slice lip is indirectly controlled by adjusting the air pad pressure which is directly controlled by adjusting the air flow valve on the high pressure side of the air re-circulation pump. Wire speed (in conjunction with the basis mass valve) is manually adjusted by the operator to control production rate.

7.4.3 Problem description

Unlike consistency control, which has been trouble-free over the years, jet : wire and pond level control have been the source of many operational problems. These problems stem from the pressure and level loops complexity in dynamics, a high degree of coupling between the two loops, significant noise in the measured values, and the dependence of the loops on external variables, such as production rate or grade, that change with time.

Man-power limitation is also a major cause of poorly tuned control loops. For instance, the duties of control engineers usually include the maintenance of the control system, the maintenance of existing computer applications, the development of new control applications, and other day-to-day "fire-fighting" assignments. Therefore, the largest problem in achieving or maintaining "healthy" control loops is the time-consuming testing required to analyze and monitor each individual control loop. From a time allocation perspective, manual loop analysis results usually fall into one of three categories: 1) a preponderance of tests achieves very good performance for hundreds of control loops, but occupies most of the control engineer's time; 2) a large number of negative results indicates that testing is insufficient, but no control engineer is available to mitigate the problems; 3) a series of time-consuming and labour-intensive tests show very good tuning results, but process drifts or changes quickly nullify the results and re-tuning is required all over again. Insufficient testing is the norm because of other time commitments and, as a result, costly sheet breaks are often the first indication of poor control performance.

The objective of this research into loop performance is to provide the plant control engineer with an on-line measure of control loop performance obviating the time-consuming manual test. This measure can be monitored on a regular basis and performance statistics can be used to schedule loop re-tuning. The result of a proactive re-tuning schedule that requires less control engineer and technician effort will maintain better overall loop performance and reduce plant downtime.

7.5 Performance assessment of the headbox control system

7.5.1 Single loop performance assessment

The schematic diagram of the current control system is shown in Figure 7.2. The present control strategy is to regulate the total head (pressure plus level) by adjusting the air pad pressure and the pond level via multiloop PID controllers. To assess control performance, we first applied the SISO loop performance assessment technique to individual loops, which may answer the following questions: 1) if the current level loop cannot be further tuned owing to some constraints, is it possible to reduce variation of the total head further by adjusting the pressure loop controller? 2) if the pressure loop is well tuned and cannot be adjusted, is it possible to reduce variation of the level further by adjusting the level loop controller?

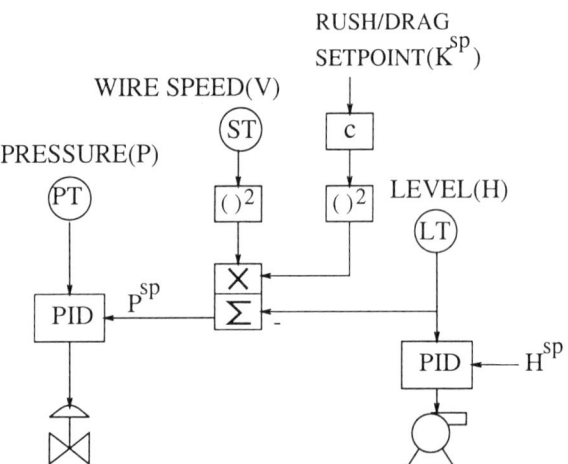

Fig. 7.2. Schematic diagram of the control system

Feedback control performance assessment of SISO processes has been discussed by Harris (1989), and Desborough and Harris (1992). The FCOR algorithm can also be applied to SISO processes directly. The setpoint of the pressure loop is adjusted randomly in this application, however, requiring a special treatment.

The process of the pressure loop can be written in a standard form as

$$p_t = q^{-d}Tu_t + Na_t$$

where T and N are (SISO) rational transfer functions. Under feedback control $u_t = -Q(p_t - p_t^{sp})$,

7.5 Performance assessment of the headbox control system

$$p_t = -q^{-d}TQ(p_t - p_t^{sp}) + Na_t$$

This is equivalent to

$$p_t - p_t^{sp} = -q^{-d}TQ(p_t - p_t^{sp}) - p_t^{sp} + Na_t$$

The random adjusted setpoint can be modelled by $p_t^{sp} = Mb_t$, where M is a rational transfer function and b_t is white noise. By defining $\epsilon_t = p_t - p_t^{sp}$, we have

$$\epsilon_t = -q^{-d}TQ\epsilon_t - Mb_t + Na_t$$

The last two terms on the right hand side of the equation may be lumped as $\theta \nu_t$ via the spectral factorization for example, where θ is a rational transfer function and ν_t is white noise. Therefore

$$\epsilon_t = -q^{-d}TQ\epsilon_t + \theta \nu_t \tag{7.13}$$

Equation 7.13 is a standard form as used in the performance assessment of SISO processes with a zero setpoint. Thus, the algorithm in Chapter 2 can be applied to assess performance of the variable ϵ_t without restriction. The only question is whether it is appropriate to replace p_t with ϵ_t as the monitored variable.

The rush : drag ratio is defined by

$$K_t \triangleq \frac{\sqrt{P_t + H_t}}{cV_t}$$

where K_t is the rush : drag ratio, V_t is the wire speed, P_t is the pressure, H_t is the level and c is a unit conversion coefficient. Here we use the notation that P_t and H_t are the original variables (actual measurements) of the pressure and the level respectively, while p_t and h_t are their deviation variables.

Ideally, the rush : drag ratio should be kept constant so that the fibre suspension can be distributed uniformly on the wire. In practice, the operator always monitors the rush : drag ratio which is an indication of how well the total head is being controlled.

It follows from Figure 7.2 that

$$P_t^{sp} = (cK^{sp}V_t)^2 - H_t \tag{7.14}$$

The setpoint of rush : drag ratio K^{sp} is set by the operator and is a constant. V_t is kept constant by the motor driving the wire. Using deviation variables, Equation 7.14 can be written as

$$p_t^{sp} = -h_t$$

Thus,

$$\epsilon_t = p_t - p_t^{sp} = p_t + h_t$$

which is a variable representing the total head variation and is indicative of the rush : drag ratio variation. This is, indeed, the most important variable for the operator to monitor continually.

At present, the performance measure of the headbox by the operating personnel is the variance of pressure and level data being sampled at intervals of 15 seconds. For this reason, the same sampling interval is used for performance assessment here. It is known from previous step tests that both the pressure and level loops have time delays of approximately three sampling intervals. Figure 7.3 shows a typical set of data which was collected over 50 hours at a 15 second sampling interval. Note that both the level and the pressure are measured by using the pressure unit, Pa. The process was under multiloop PID control with a control sampling interval of 1 second. Applying the FCOR algorithm to variables ϵ_t and h_t respectively yields results shown in Figure 7.4, where each performance index is calculated from 1000 data points, *i.e.*, 4.2 hours for calculation of each performance index.

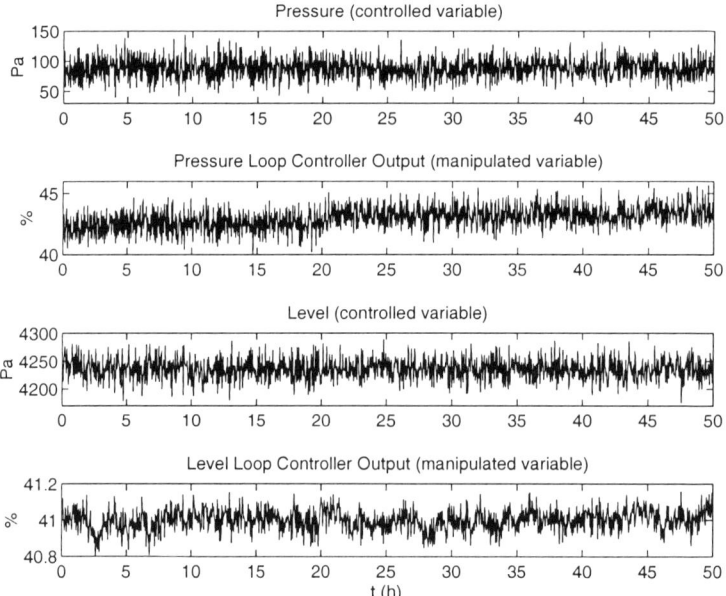

Fig. 7.3. Process data trajectory

By definition, the performance index should be in-between '0' and '1'. While '1' means the best performance, '0' means the worst performance including unstable control. For example, a performance index of 0.5 implies that current variance can be reduced (potentially) by a factor of 0.5 if an optimal tuning is implemented. Depending on the application, the loop performance measure can be classified as optimal/good/bad or acceptable/unacceptable.

7.5 Performance assessment of the headbox control system   77

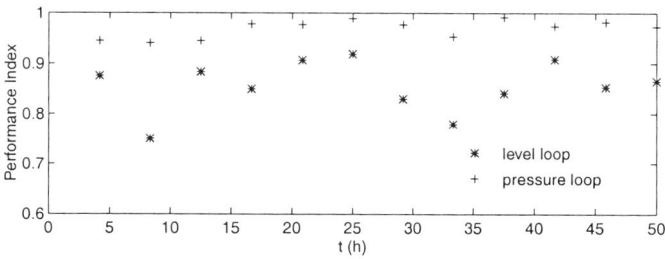

Fig. 7.4. Performance assessment from the single-input and single-output approach.

Single loop performance assessment results shown in Figure 7.4 clearly indicate an 'optimal' performance of current single loop tunings of both loops. Note that the benchmark performance is minimum variance control of the individual loop; therefore, there is little potential for further reduction in process variance by adjusting or re-designing the controller individually. It should be noted that the performance index is cycling periodically owing to differences in ambient temperature between day-and-night. Although the headbox process is housed inside a building, ambient conditions do affect the pulp quality before it enters the headbox.

7.5.2 Multivariate performance assessment

SISO performance assessment can only indicate the potential for performance improvement by adjusting individual loops. Since the level and the pressure loops are coupled, a multivariate control strategy can further reduce process variations. Multivariate performance assessment can provide the measure of such potential.

Previous tests provided the following *a priori* knowledge of the process: 1) the time delay from a change of the air flow control valve to both the pressure and level responses is approximately three sampling intervals; 2) the time delay from the fan pump speed change to the level response is approximately three sampling intervals, and 3) the time delay from the fan pump speed change to the pressure response is approximately two sampling intervals. It follows from Tsiligiannis and Svoronos (1988) that this process has a diagonal interactor matrix. For example, by examining each row of the transfer function matrix, it can be shown that each output can be paired to an input with the minimum delay in that row. For detailed discussion of the occurrence of diagonal interactor matrices in practice, readers are referred to Appendix B.

Since the pressure loop has a randomly adjusted setpoint and the variable of interest in this assessment is the total head $(p+h)$, the problem should be reformulated as it was in the SISO case. The multivariable model with the outputs p_t and h_t can be written as

$$p_t = T_{11}u_1 + T_{12}u_2 + N_{11}a_1 + N_{12}a_2 \tag{7.15}$$
$$h_t = T_{21}u_1 + T_{22}u_2 + N_{21}a_1 + N_{22}a_2 \tag{7.16}$$

where u_1 is the manipulated variable of the air flow control valve; u_2 is the manipulated variable of the fan pump speed; T_{11}, T_{21} and T_{22} are approximately first-order transfer functions while T_{12} can be approximated by a second order transfer function; N_{ij} (for $i = 1, 2, j = 1, 2$) are disturbance transfer functions and can be modelled by autoregressive moving average processes. The pressure, p_t, is measured by a low range differential pressure transmitter with the low side of the transmitter vented to the atmosphere. The level, h_t, is measured by a flange mounted transmitter. The air flow control valve on the suction of the vacuum pump is a Fisher V100 segmented ball valve. The fan pump speed is controlled by a variable speed DC drive. Multiloop PID controllers are implemented in the process. The pressure loop is controlled by manipulating the air flow control valve, and the level loop is controlled by manipulating the fan pump speed. Note that, except for a *priori* knowledge of the time-delays in T_{ij}, other information about T_{ij}, N_{ij} and controller transfer functions is not required for performance assessment.

It follows from the analysis in the SISO case that (in the deviation variable sense)

$$p_t^{sp} = -h_t \tag{7.17}$$

Substituting Equation 7.16 into 7.17 yields

$$p_t^{sp} = -T_{21}u_1 - T_{22}u_2 - N_{21}a_1 - N_{22}a_2 \tag{7.18}$$

Subtracting Equation 7.18 from Equation 7.15 yields

$$p_t - p_t^{sp} = (T_{11}+T_{21})u_1+(T_{12}+T_{22})u_2+(N_{11}+N_{21})a_1+(N_{12}+N_{22})a_2 \tag{7.19}$$

For simplicity, Equation 7.19 can be written as

$$\epsilon_t = T'_{11}u_1 + T'_{12}u_2 + N'_{11}a_1 + N'_{12}a_2 \tag{7.20}$$

It follows from Equation 7.17 that $\epsilon_t = p_t - p_t^{sp} = p_t + h_t =$ "total head", which is, conveniently, the variable of interest that we would like to monitor. More importantly, Equation 7.20 represents the pressure equation, which transfers the random setpoint to a constant (zero) setpoint; therefore, Equation 7.20 together with 7.16 represents a 2×2 multivariable process with constant setpoints for both variables ϵ_t and h_t. In this case, it can be easily verified that the time delay structure of the multivariable process with process variables ϵ_t and h_t is the same as that of the multivariable process with process variables p_t and h_t. Therefore, performance assessment methods, as introduced in the foregoing sections, can readily be applied and their results are shown in Figure 7.5.

7.5 Performance assessment of the headbox control system

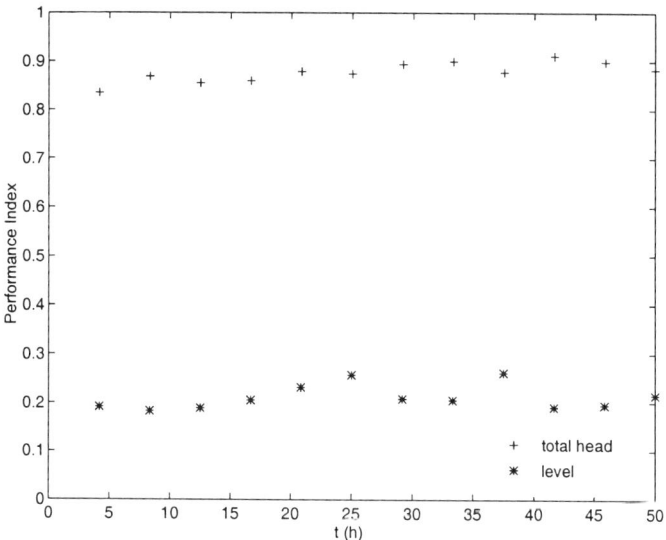

Fig. 7.5. Performance assessment from the multivariate approach.

In Figure 7.5, '+' represents the performance index of the total head, and '*' represents the performance index of the level output. The average performance index of the total head is 0.85 and the average performance index of the level is 0.22. Note that the benchmark is multivariable minimum variance control. Compared to the SISO assessment which yields the performance index of the total head as an almost perfect '1', multivariate assessment does indicate that, if desired, there is potential to reduce the variation of the total head very slightly by implementing multivariable control. However, the improvement may not be significant enough to justify implementation of the multivariable control or an interaction compensator in practice. Clearly, there is a potential to reduce level variation by implementing multivariate control. Reduction of level variation is not the objective, however, so in order to reduce the variance of the total head, alternative control strategies, such as feedforward control or reduction of the dead time, may be necessary if further improvement in performance is desired.

Since the current total head variance is 260.6 Pa^2 (0.0042 [in. $H_2O]^2$) and level variance is 387 Pa^2 (0.0062 [in. $H_2O]^2$), the multivariable performance indices also imply that within the present control framework, the total head variance is never less than $0.85 \times 260.6 = 221.5$ Pa^2 (0.0036 [in. $H_2O]^2$), and the level variance is never less than $0.22 \times 387.7 = 85.3$ Pa^2 (0.0014 [in. $H_2O]^2$).

In this way the assessment results indicate that

1. The overall control performance is subject to periodic changes owing to the difference between day and night ambient conditions;
2. Within the framework of the single-loop controller tuning, there is little potential either to reduce the total-head variance by tuning the pressure controller or to reduce the level variance by tuning the level controller;
3. Within the current control structure, total-head variance is no less than 221.5 Pa^2 (0.0036 [in. H$_2$O]^2) under any linear (SISO or MIMO) feedback control;
4. The present controllers have been well tuned, and it may be unnecessary to adjust controller tunings further or implement a multivariable control algorithm;
5. Since any possible deduction in the total-head variance by tuning or redesigning the feedback control is no more than 15%, it may be necessary to implement feedforward control or to reduce dead times or to change the current control structure in order to reduce the variance of the total head significantly.

Since the selected data sampling frequency is lower than the controller sampling frequency in this example, as stated in the previous sections, a possible underestimation of the performance indices may be expected. However, any possible underestimation error only implies that the actual performance is even better than the estimated performance or the actual lower bound is larger than the estimated lower bound; therefore, it does not affect any of the above conclusions.

7.6 Summary

Control loop performance assessment of multivariate processes with diagonal interactor matrices has been introduced. Both single loop performance and multivariable performance of the headbox control loops have been estimated from actual industrial data. The results have shown that the present controllers have been well tuned, and it is unnecessary to adjust controller tunings further or to implement multivariate control under the current control structure. In order to improve control performance further, the implementation of feedforward control, reduction of dead times or redesign of the control structure may be necessary; therefore, these results provide guidelines for control loop tuning and provide an insight for the potential benefits of exploring multivariable control strategy.

CHAPTER 8
FEEDBACK CONTROLLER PERFORMANCE ASSESSMENT: GENERAL INTERACTOR

8.1 Introduction

The factorization of a simple or diagonal interactor matrix only requires *a priori* knowledge of the pure time-delays of each element in a transfer function matrix. The factorization of a general interactor matrix, however, requires a complete knowledge of, or, at least, the first few Markov parameter matrices of a MIMO process. The *unitary interactor matrix*, as introduced by Peng and Kinnaert(1992), plays an important role in feedback controller performance assessment of processes with general interactor matrices. The estimation of the unitary interactor matrix using closed-loop data from simple closed-loop tests has been discussed in Chapter 5.

The main contributions of this chapter are to obtain the feedback controller-invariant term for MIMO processes with general interactor matrices and to propose a control loop performance measure that is conceptually simple and computationally efficient. The algorithm is valid for the performance assessment of all classes of multivariable (square or non-square) systems. Although the method introduced in this chapter can be applied to simple or diagonal interactor matrices as well, it is not recommended since the algorithms introduced in Chapters 6 and 7 are more efficient in handling these special classes of processes.

This chapter is organized as follows: First, the interactor matrix is reviewed, and a suitable expression for the MIMO feedback controller-invariant term, which is the measure of the minimum variance control loop performance, is derived in Section 8.2. The key ingredient of this scheme is filtering and subsequent correlation analysis. The FCOR algorithm is used to estimate the achievable multivariate minimum variance performance from routine operating data. Its derivation for the general interactor matrix is considered in Section 8.3. The application of the FCOR algorithm to a simulated square and a non-square MIMO process, and to an industrial absorption unit is considered in Section 8.4, followed by concluding remarks in Section 8.5.

8.2 Feedback controller-invariance of minimum variance term

8.2.1 Review of the unitary interactor matrix

Consider the MIMO process with a general interactor matrix:

$$Y_t = TU_t + Na_t \qquad (8.1)$$

where T and N are proper (causal), rational transfer function matrices in the backshift operator q^{-1}; Y_t, U_t, and a_t are output, input and white-noise vectors of appropriate dimensions.

Wolovich and Falb (1976) and Goodwin and Sin (1984) have shown the existence of a unique lower triangular form of the general interactor matrix. However, the interactor matrix can also take other forms. It can be a full matrix or an upper triangular matrix (Shah, Mohtadi, and Clarke 1987). Rogozinski et al. (1987) have introduced an algorithm for the calculation of a *nilpotent interactor matrix*. Peng and Kinnaert (1992) have introduced the *unitary interactor matrix*, which is a special case of the nilpotent interactor matrix. The unitary interactor matrix has been discussed in Chapter 3. Some important properties of the unitary interactor matrix in minimum variance or singular LQ control have been discussed in Chapter 4.

The existence of the unitary interactor matrix is established in Peng and Kinnaert (1992). The unitary interactor matrix has been shown to be an "ideal" factorization of the time delay matrix for minimum variance or singular LQ type control in Chapter 4. A simple algorithm exists for the calculation of the unitary interactor matrix (Rogozinski, Paplinski, and Gibbard 1987; Peng and Kinnaert 1992) (also see Appendix A. The traditional procedure for factorization of the general interactor matrix does require complete knowledge of the transfer function matrix (Wolovich and Falb 1976; Goodwin and Sin 1984; Rogozinski, Paplinski, and Gibbard 1987; Peng and Kinnaert 1992). In Chapter 5, it has been shown that the factorization of the interactor matrix can be achieved from the first few Markov parameter or impulse response coefficient matrices of the process. Consequently, the estimation of the unitary interactor matrix is simplified and can be achieved by using closed-loop data from simple closed-loop tests. The two special interactor matrices, the *simple* interactor matrix, $D = q^{-d}I$, and the *diagonal* interactor matrix, $D = diag(q^{-d_1}, \cdots, q^{-d_n})$, are also unitary interactor matrices. Performance assessment for the simple interactor and the diagonal interactor has been discussed in Chapters 6 and 7 respectively.

8.2.2 Feedback controller-invariance of minimum variance term and its separation from routine operating data

Minimum variance control (or the benchmark control) is defined by a feedback control law which minimizes the output LQ objective function without

8.2 Feedback controller-invariance of minimum variance term

penalty to the control action:
$$J = E(Y_t - Y_t^{sp})^T(Y_t - Y_t^{sp})$$
or the weighted LQ objective function:
$$J = E(Y_t - Y_t^{sp})^T W (Y_t - Y_t^{sp})$$

This is also regarded as Singular LQ control (Peng and Kinnaert 1992). For simplicity in the following proof, we shall first assume that the setpoint, Y_t^{sp}, is zero and the weighting function $W = I$. Then the singular LQ objective function is reduced to
$$J = E[Y_t^T Y_t]$$
The general case is discussed in Remark 8.2.2 when the setpoint is not zero, and in Remark 8.2.4 when $W \neq I$.

Consider the multivariable process
$$Y_t = TU_t + Na_t$$

Using the notation of multivariable minimum variance control due to Goodwin and Sin (1984), the minimum variance control law can be designed to make the variance of the output DY_t or, equivalently, $\tilde{Y}_t = q^{-d}DY_t$ minimum, where the positive integer d is the maximum order (highest power of q) of all the elements of the interactor matrix, D. The filter, $q^{-d}D$, removes infinite zeros from the transfer function matrix. Since D is a unitary interactor matrix, the singular LQ or minimum variance control laws for $\tilde{Y}_t$ and Y_t are the same. This important property of the unitary interactor matrix is discussed in Remark 8.2.3. Unlike previous work on the design of multivariable minimum variance control by Goodwin and Sin(1984) and many others in the literature (Tsiligiannis and Svoronos 1988; Harris and MacGregor 1987), the main focus of this study is the derivation of a suitable expression for the feedback controller-invariant, minimum variance term from routine operating data.

Theorem 8.2.1. *For a multivariable process*
$$Y_t = TU_t + Na_t \tag{8.2}$$
the minimum variance control is obtained by minimizing
$$J = E[\tilde{Y}_t^T \tilde{Y}_t] \tag{8.3}$$
where $\tilde{Y}_t = q^{-d}DY_t$ is the interactor-filtered output. The performance measure is then given by the following steps:

1. The quadratic measure of minimum variance is given by

$$E[\tilde{Y}_t^T \tilde{Y}_t]_{min} = E(e_t^T)(e_t) = tr(Var(Fa_t))$$

where $e_t = Fa_t$, the polynomial matrix F depends only on the interactor matrix and the noise model, and satisfies the identity:

$$q^{-d}DN = \underbrace{F_0 + \cdots + F_{d-1}q^{-(d-1)}}_{F} + q^{-d}R \quad (8.4)$$

where R is a proper rational transfer function matrix;

2. If one models closed-loop routine operating data under feedback control by the following multivariate moving-average process:

$$\tilde{Y}_t - E(\tilde{Y}_t) = \underbrace{F_0 a_t + F_1 a_{t-1} + \cdots + F_{d-1} a_{t-d+1}}_{e_t}$$

$$+ \underbrace{L_0 a_{t-d} + L_1 a_{t-d-1} + \cdots}_{w_{t-d}} \quad (8.5)$$

then the minimum variance term, $e_t = Fa_t$, consists of the first d terms of this moving-average model and can be separated from time series analysis of routine operating data and used as a benchmark measure of multivariate minimum variance control, therefore.

Proof. Recall that in Chapter 6, for a process with a *simple interactor matrix*, i.e., $D = q^{-d}I$,

$$Y_t = q^{-d}\tilde{T}U_t + Na_t \quad (8.6)$$

where $\tilde{T}$ is the delay-free transfer function matrix, the following inequality holds:

$$E[Y_t^T Y_t] \geq E(e_t^T)(e_t) = tr(Var(Fa_t))$$

where $e_t = Fa_t$, and F is defined by the identity

$$N = \underbrace{F_0 + F_1 q^{-1} + \cdots + F_{d-1} q^{-(d-1)}}_{F} + q^{-d}R$$

where F_i, for $i = 0, \cdots, d-1$, are constant coefficient matrices, and R is a proper rational transfer function matrix. The equality holds when the minimum variance control law is implemented on the process.

If routine closed-loop operating data under feedback control is modelled by a multivariate moving-average process:

$$Y_t - E(Y_t) = \underbrace{F_0 a_t + F_1 a_{t-1} + \cdots + F_{d-1} a_{t-d+1}}_{e_t} + \underbrace{L_0 a_{t-d} + L_1 a_{t-d-1} + \cdots}_{w_{t-d}}$$

$$(8.7)$$

then the term w_{t-d} is feedback controller-dependent, and the term e_t, consisting of the first d terms of the moving-average model, is independent of

8.2 Feedback controller-invariance of minimum variance term

feedback control. Under minimum variance control, w_{t-d} vanishes and, therefore, e_t represents the minimum variance term and can be separated from time series analysis of routine operating data.

Now consider the process with a general *unitary interactor matrix* i.e., $D \neq q^{-d}I$:
$$Y_t = TU_t + Na_t = D^{-1}\tilde{T}U_t + Na_t \tag{8.8}$$
Multiplying both sides of Equation 8.8 by $q^{-d}D$ yields
$$\begin{aligned} q^{-d}DY_t &= q^{-d}\tilde{T}U_t + q^{-d}DNa_t \\ &= q^{-d}\tilde{T}U_t + \tilde{N}a_t \end{aligned} \tag{8.9}$$
where $\tilde{N}$ is a proper transfer function matrix. By defining $\tilde{Y}_t = q^{-d}DY_t$, Equation 8.9 is now transformed to the same form as 8.6, *i.e.*,
$$\tilde{Y}_t = q^{-d}\tilde{T}U_t + \tilde{N}a_t \tag{8.10}$$
This is a process with a *simple interactor matrix*. It follows that
$$E[\tilde{Y}_t^T \tilde{Y}_t] \geq E(\tilde{e}_t^T)(\tilde{e}_t) = tr(Var(\tilde{F}a_t))$$
where $\tilde{e}_t = \tilde{F}a_t$, $\tilde{F}$ is defined by the identity:
$$\tilde{N} = q^{-d}DN = \underbrace{\tilde{F}_0 + \tilde{F}_1 q^{-1} + \cdots + \tilde{F}_{d-1} q^{-(d-1)}}_{\tilde{F}} + q^{-d}\tilde{R} \tag{8.11}$$

Thus, if the interactor-filtered, routine operating data under feedback control is modelled by a multivariate moving-average process:
$$\tilde{Y}_t - E(\tilde{Y}_t) = \underbrace{\tilde{F}_0 a_t + \tilde{F}_1 a_{t-1} + \cdots + \tilde{F}_{d-1} a_{t-d+1}}_{\tilde{e}_t} + \underbrace{\tilde{L}_0 a_{t-d} + \tilde{L}_1 a_{t-d-1} + \cdots}_{\tilde{w}_{t-d}}$$
$$\tag{8.12}$$
then the lower bound term $\tilde{e}_t$ is independent of feedback control and can be estimated from routine operating data, therefore. To simplify notation, the tilde signs, $\tilde{\ }$, on the right hand side of Equations 8.12 and 8.11 have been dropped in the statement of Theorem 8.2.1.

Remark 8.2.1. The feedback controller-invariant property of the minimum variance term is valid for both square and non-square transfer function matrices. When the transfer function is of full rank with the row dimension smaller than the column dimension, then the minimum variance is achievable. This is clear from the proofs in Chapter 4 and Chapter 6, where the inverse of $\tilde{T}$, $\tilde{T}^{-1}$, is replaced by its pseudoinverse, $\tilde{T}^\dagger$. On the other hand, when the transfer function matrix is a non-square matrix with the row dimension larger than the column dimension, then the feedback controller-invariant term may not be achievable. Since a unitary interactor matrix can always be factored irrespective of whether the transfer function matrix is a square or non-square matrix (Rogozinski, Paplinski, and Gibbard 1987), the methodology for performance assessment as proposed in this chapter is valid for both square and non-square transfer function matrices.

Remark 8.2.2. If the setpoint is not zero, then we define

$$\epsilon_t \triangleq Y_t^{sp} - Y_t$$

The interactor-filtered singular LQ objective is now written as

$$J = E(\tilde{Y}_t - \tilde{Y}_t^{sp})^T(\tilde{Y}_t - \tilde{Y}_t^{sp}) = E[\tilde{\epsilon}_t^T \tilde{\epsilon}_t]$$

where $\tilde{Y}_t = q^{-d}DY_t$, $\tilde{Y}_t^{sp} = q^{-d}DY_t^{sp}$ and $\tilde{\epsilon}_t = q^{-d}D\epsilon_t$. Chapter 11 will show that in this case, instead of using $\tilde{Y}_t$ for the time series analysis as shown in Theorem 8.2.1, $\tilde{\epsilon}_t$ should be used for the analysis. Then the first d terms of the moving-average model of $\tilde{\epsilon}_t$ constitute the feedback controller-invariant minimum variance term.

Remark 8.2.3. It has been shown in Chapter 4 that if D is a unitary interactor matrix, then the minimum variance control law, which minimizes the following objective function of the interactor-filtered variable $\tilde{Y}_t$:

$$J_1 = E(\tilde{Y}_t - \tilde{Y}_t^{sp})^T(\tilde{Y}_t - \tilde{Y}_t^{sp}) \tag{8.13}$$

also minimizes the objective function of the original variable Y_t:

$$J_2 = E(Y_t - Y_t^{sp})^T(Y_t - Y_t^{sp}) \tag{8.14}$$

and $J_1 = J_2$.

Thus, the performance measure of the original variable Y_t can be obtained via the performance measure of the interactor-filtered variable $\tilde{Y}_t$.

Remark 8.2.4. In Remark 8.2.3, if the unitary interactor matrix is replaced by a weighted unitary interactor matrix, D_w, then

$$E(\tilde{Y}_t - \tilde{Y}_t^{sp})^T(\tilde{Y}_t - \tilde{Y}_t^{sp}) = E(Y_t - Y_t^{sp})^T W(Y_t - Y_t^{sp})$$

where D_w satisfies $D_w^T D_w = W$. The existence and factorization of such a weighted unitary interactor matrix can be found in Chapter 4. With such an interactor matrix, the minimum variance control law for the interactor-filtered variable, $\tilde{Y}$, is identical to the weighted minimum variance control law for the original variable, Y_t. In fact, in Chapter 4, it has been shown that the minimum variance control law is identical to the singular LQ control law solved via the spectral factorization method (Harris and MacGregor 1987).

To summarize Theorem 8.2.1 and Remarks 8.2.1 to 8.2.4, the following general result is presented:
For a multivariable (square/non-square) process

$$Y_t = TU_t + Na_t$$

the minimum variance control is obtained by minimizing

$$J = E[(Y_t - Y_t^{sp})^T W(Y_t - Y_t^{sp})] = E(\epsilon_t^T W \epsilon_t)$$

If one models closed-loop routine operating data under feedback control by the following multivariate moving-average process:

$$\tilde{\epsilon}_t - E(\tilde{\epsilon}_t) = \underbrace{F_0 a_t + F_1 a_{t-1} + \cdots + F_{d-1} a_{t-d+1}}_{e_t}$$
$$+ \underbrace{L_0 a_{t-d} + L_1 a_{t-d-1} + \cdots}_{w_{t-d}}$$

where $\tilde{\epsilon}_t = q^{-d} D_w \epsilon_t$, then the minimum variance term, $\tilde{\epsilon}_t|_{mv} = e_t = F a_t$, consists of the first d terms of this moving-average model, and can be separated from time series analysis of routine operating data, therefore. The quadratic benchmark performance measure, $E(\epsilon_t^T W \epsilon_t)|_{mv}$, can then be calculated from $E(e_t^T e_t)$.

Remark 8.2.5. Deduction of the minimum variance control benchmark performance as in this chapter requires knowledge of only the time-delays (or infinite zeros) of the transfer function matrix. No other information is required. In practice, there are many limitations in reducing output variance through feedback control. Control action constraints, the existence of poorly damped or unstable (NMP) zeros, desired robustness characteristics *etc.*, are examples of such limitations. Time-delays or the interactor matrices are the most fundamental level of limitation in reducing variance and are the only possible performance limitations that can be estimated from routine operating data. The identification of such benchmark performance does not imply implementation of such a control law, but this benchmark performance provides an *absolute* lower bound or *global* minimum (Astrom and Wittenmark 1990) of process variance and "so can be used much like the Cramer-Rao lower bound on variance in statistical parameter estimation (Harris, Boudreau, and MacGregor 1996)". Even for minimum phase systems, the implementation of minimum variance control is usually not recommended (Desborough and Harris 1992; Eriksson and Isaksson 1994) owing to its poor robustness and need for excessive control action. Certainly for non-minimum phase systems, it would be imprudent to implement such a benchmark control law. Nevertheless, minimum variance control does serve as a useful "global minimum" reference point and provides a first-level benchmark against which to assess current control performance. This first-level performance measure is obtained with minimum effort—from routine operating data together with *a priori* knowledge of the time delay matrix. This *a priori* knowledge can be obtained from simple tests of the MIMO process under closed-loop conditions and subsequent SVD analysis of the data as discussed in Chapter 5. Using this SVD method, entire knowledge of the model of the process is not a necessary prerequisite for estimating the interactor. Thus, the proposed method provides an efficient tool to monitor modern processing facilities which can have hundreds and possibly thousands of control loops comprehensively. For those loops which indicate good first-level performance measures, no further adjustment or testing is necessary. For loops which indicate poor performance measures, a second-level study, which may require process identification and/or redesign or retuning of control loops, may be necessary. Thus, the second-level performance assessment need only be conducted on a limited number of loops and this saves valuable personnel time. The benchmark performance (*local* minimum), subject to constraints such as non-minimum phase and control action constraints, can then be studied at this second-level performance assessment; other performance measures such as regulation of step-type disturbances (Eriksson and Isaksson 1994) are also studied at this level. These issues will be discussed in the following chapters.

8.3 The FCOR algorithm for general interactor matrices

8.3.1 Multivariable performance measures

As proved in the previous sections, performance assessment of multivariable processes can be reduced to finding the minimum variance term, e_t, from a multivariate moving average process, which has the general form shown in Equation 8.5. From Equation 8.5, the covariance between the output and the white noise sequence at lag i (for $i < d$) is given by

$$E[\tilde{Y}_t a_{t-i}^T] = F_i \Sigma_a \triangleq \Sigma_{\tilde{Y}a}(i) \tag{8.15}$$

where $\Sigma_a = E(a_t a_t^T)$. From

$$q^{-d} D Y_t|_{mv} \triangleq \tilde{Y}_t|_{mv} = e_t = F_0 a_t + \cdots + F_{d-1} a_{t-d+1}$$

one can solve for $Y_t|_{mv}$ as

$$Y_t|_{mv} = q^d D^{-1}(F_0 a_t + F_1 a_{t-1} + \cdots + F_{d-1} a_{t-d+1})$$

where $\tilde{Y}_t|_{mv}$ is the interactor-filtered output under minimum variance control, and $Y_t|_{mv}$ is the original output under the same control law. Note that from Remark 8.2.3 the minimum variance control laws of Y_t and $\tilde{Y}_t$ are identical. For the unitary interactor matrix, we have $(D^{-1}(q) = D^T(q^{-1}))$, i.e.,

$$\begin{aligned} D^{-1} &= (D_0 q^d + \cdots + D_{d-1} q)^{-1} \\ &= D_0^T q^{-d} + \cdots + D_{d-1}^T q^{-1} \end{aligned} \tag{8.16}$$

Therefore

$$\begin{aligned} Y_t|_{mv} &= (D_0^T + \cdots + D_{d-1}^T q^{d-1})(F_0 + \cdots + F_{d-1} q^{-d+1}) a_t \\ &\triangleq (E_0 + E_1 q^{-1} + \cdots + E_{d-1} q^{-d+1}) a_t \end{aligned} \tag{8.17}$$

8.3 The FCOR algorithm for general interactor matrices

(Note that for a weighted unitary interactor matrix, $D_w^{-1}(q) \neq D_w^T(q^{-1})$, but $D_w = D_{uni}W^{1/2}$ and D_{uni} is a unitary interactor. Thus, $D_w^{-1} = W^{-1/2}D_{uni}^T(q^{-1})$ and Equation 8.16 has to be modified accordingly.)

Owing to causality, any term with a positive power of q in Equation 8.17 must be zero. Equation 8.17 can be written as a compact matrix form:

$$[E_0, E_1, \cdots, E_{d-1}] = \\ [D_0^T, D_1^T, \cdots, D_{d-1}^T] \begin{bmatrix} F_0 & F_1 & \cdots & F_{d-1} \\ F_1 & F_2 & \cdots & \\ \vdots & \vdots & & \\ \vdots & & & F_{d-1} \\ F_{d-1} & & & \end{bmatrix} \quad (8.18)$$

From Equation 8.17, variance of Y_t under minimum variance control can be written as

$$\Sigma_{mv} = Var(Y_t|mv) = E_0 \Sigma_a E_0^T + \cdots + E_{d-1}\Sigma_a E_{d-1}^T$$
$$\stackrel{\triangle}{=} XX^T \quad (8.19)$$

$$\text{where} \quad X \stackrel{\triangle}{=} [E_0 \Sigma_a^{1/2}, E_1 \Sigma_a^{1/2}, \cdots, E_{d-1}\Sigma_a^{1/2}] \quad (8.20)$$

From Equation 8.15, we have

$$F_i = \Sigma_{\tilde{Y}_a}(i)\Sigma_a^{-1} \quad (8.21)$$

Substituting Equation 8.21 into 8.18, and then substituting the result into 8.20 yields

$$X = [D_0^T, D_1^T, \cdots, D_{d-1}^T] \times \\ \begin{bmatrix} \Sigma_{\tilde{Y}_a}(0)\Sigma_a^{-1/2} & \Sigma_{\tilde{Y}_a}(1)\Sigma_a^{-1/2} & \cdots & \Sigma_{\tilde{Y}_a}(d-1)\Sigma_a^{-1/2} \\ \Sigma_{\tilde{Y}_a}(1)\Sigma_a^{-1/2} & \Sigma_{\tilde{Y}_a}(2)\Sigma_a^{-1/2} & \cdots & \\ \vdots & \vdots & & \\ & & & \Sigma_{\tilde{Y}_a}(d-1)\Sigma_a^{-1/2} \\ \Sigma_{\tilde{Y}_a}(d-1)\Sigma_a^{-1/2} & & & \end{bmatrix}$$

Since variance of Y_t under minimum variance control can be calculated from Equation 8.19, the objective function based performance measure (denoted as MIMO performance measure) can be calculated as

$$\eta(d) = \frac{\text{minimum variance}}{\text{actual variance}} = \frac{E[Y_t^T Y_t]_{min}}{E[Y_t^T Y_t]}$$
$$= \frac{tr\Sigma_{mv}}{tr(E[Y_t Y_t^T])}$$
$$= \frac{tr(XX^T)}{tr\Sigma_Y}$$

It is often desirable to compare the variance-covariance matrix of the actual output with the variance-covariance matrix of the ideal output under minimum variance control. The performance indices of individual outputs are obtained from the diagonal elements of such a comparison:

$$[\eta_{y_1}, \cdots, \eta_{y_n}]^T = diag\{\Sigma_{mv}\tilde{\Sigma}_Y^{-1}\} = diag\{XX^T\tilde{\Sigma}_Y^{-1}\}$$

where $\tilde{\Sigma}_Y = diag(\Sigma_Y)$. The individual output performance indices represent the performance of each output with respect to the ideal output under multivariable minimum variance control. If an offset exists in the process output, then the output variance, Σ_Y, should be replaced by the output mean square error in the above calculation of the performance indices.

Although a_t is unknown in this calculation, it can be replaced by the estimated "white" noise sequence, $\hat{a}_t$, or the innovation term via time series analysis as introduced in Section 2.3.2. This whole procedure of obtaining the multivariable performance index is the FCOR algorithm. The FCOR approach provides a relatively easy way to calculate the performance measure of a multivariable process.

Harris et al. (1996) have proposed another approach to assess the performance of multivariable systems using spectral factorization to normalize the lower triangular interactor matrix and subsequently using the Diophantine identity to calculate the benchmark performance. Readers are referred to Harris et al. (1996) for detailed discussion.

8.4 Evaluation of the FCOR algorithm on a simulation and an industrial application

8.4.1 Simulated example

Example 8.4.1. This simulation example demonstrates the application of the FCOR algorithm to both square and non-square multivariable systems with the general interactor matrix.

Consider a 2×2 multivariable process with the open-loop transfer function matrix T and disturbance transfer function matrix N given by

$$T = \begin{bmatrix} \frac{q^{-1}}{1-0.4q^{-1}} & \frac{K_{12}q^{-2}}{1-0.1q^{-1}} \\ \frac{0.3q^{-1}}{1-0.1q^{-1}} & \frac{q^{-2}}{1-0.8q^{-1}} \end{bmatrix}$$

$$N = \begin{bmatrix} \frac{1}{1-0.5q^{-1}} & \frac{-0.6}{1-0.5q^{-1}} \\ \frac{0.5}{1-0.5q^{-1}} & \frac{1.0}{1-0.5q^{-1}} \end{bmatrix}$$

The white noise excitation, a_t, is a two-dimensional normally-distributed white noise sequence with $\Sigma_a = I$. The output quality is measured by $J = E[Y_t^T Y_t]$. A unitary interactor matrix D can then be factored out as

8.4 Evaluation of the FCOR algorithm: simulation and application

$$D = \begin{bmatrix} -0.9578q & -0.2873q \\ -0.2873q^2 & 0.9578q^2 \end{bmatrix} \tag{8.22}$$

Then DN is given by

$$DN = \begin{bmatrix} \frac{-1.1014q}{(1-0.5q^{-1})} & \frac{0.2874q}{(1-0.5q^{-1})} \\ \frac{0.1916q^2}{(1-0.5q^{-1})} & \frac{1.1302q^2}{(1-0.5q^{-1})} \end{bmatrix}$$

$q^{-d}DN$ ($d = 2$) can be separated in the form of Equation 8.4, where F and R matrices are obtained as

$$F = \begin{bmatrix} -1.1014q^{-1} & 0.2874q^{-1} \\ 0.1916 + 0.0958q^{-1} & 1.1302 + 0.5651q^{-1} \end{bmatrix} \tag{8.23}$$

$$R = \begin{bmatrix} \frac{-0.5507}{1-0.5q^{-1}} & \frac{0.1437}{1-0.5q^{-1}} \\ \frac{0.0479}{1-0.5q^{-1}} & \frac{0.2826}{1-0.5q^{-1}} \end{bmatrix} \tag{8.24}$$

The feedback controller-invariant term is, therefore,

$$e_t = Fa_t = \begin{bmatrix} -1.1014q^{-1} & 0.2874q^{-1} \\ 0.1916 + 0.0958q^{-1} & 1.1302 + 0.5651q^{-1} \end{bmatrix} a_t$$

The theoretical minimum variance matrix under the minimum variance control can be calculated from $Y_t|_{mv} = q^d D^{-1} F a_t$ as

$$Y_t|_{mv} = \begin{bmatrix} 1 - 0.02752q^{-1} & -0.6 - 0.1623q^{-1} \\ 0.5 + 0.09176q^{-1} & 1 + 0.5412q^{-1} \end{bmatrix} a_t$$

This will be the theoretical benchmark to assess performance of the feedback controller. The FCOR algorithm will estimate this benchmark from routine operating data.

Consider the following multiloop minimum variance controller based on the two single loops without interaction compensation:

$$Q = \begin{bmatrix} \frac{0.5-0.20q^{-1}}{1-0.5q^{-1}} & 0 \\ 0 & \frac{0.25-0.200q^{-1}}{(1-0.5q^{-1})(1+0.5q^{-1})} \end{bmatrix}$$

The *a priori* knowledge of the interactor matrix can be estimated either from previous open-loop tests or from simple closed-loop tests. One can then apply the FCOR algorithm to the interactor-filtered variable $\tilde{Y}_t$, and the MIMO performance index of the original variable Y_t can be estimated. The result is shown in Figure 8.1, where performance indices include an objective function-based performance index (denoted as MIMO) and individual output performance indices (denoted as y1 and y2 respectively). In this example, when $K_{12} \to 0$ (K_{12} is the numerator gain of element (1,2) of the process transfer function matrix T.), the performance measures of both outputs reach the value '1' owing to weak interaction. However, with increasing interaction (*i.e.*, as K_{12} increases) the performance deteriorates, and eventually the performance index of y_1 approaches zero. Performance of y_1 is more sensitive to the change in K_{12}. It appears that the objective function based MIMO performance index is influenced significantly by y_1 in this example. This plot also shows a good agreement of the estimated performance indices with the theoretical ones.

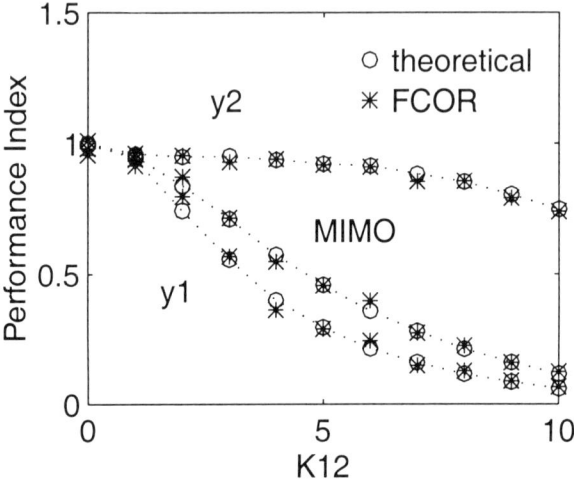

Fig. 8.1. Performance assessment of a square MIMO process (with a general interactor matrix) under multiloop minimum variance control

To see the effect of output weighting, a weighting matrix

$$W = \begin{bmatrix} 1 & 0 \\ 0 & 4 \end{bmatrix}$$

is assumed, *i.e.*, y_2 is regarded as a more important output variable than y_1. Following the procedure in Chapter 4, a weighted unitary interactor matrix is formed as

$$D_w = \begin{bmatrix} -0.8575q & -1.029q \\ 0.5145q^2 & -1.715q^2 \end{bmatrix}$$

The feedback control invariant polynomial matrix F can be calculated from $q^{-d}D_w N = F + q^{-d}R$ as

$$F = \begin{bmatrix} -1.3720q^{-1} & -0.5145q^{-1} \\ -0.3430 - 0.1715q^{-1} & -2.0240 - 1.0120q^{-1} \end{bmatrix}$$

The theoretical minimum variance matrix under the weighted minimum variance control can be calculated from $Y_t|_{mv} = q^d D_w^{-1} F a_t$ as

8.4 Evaluation of the FCOR algorithm: simulation and application 93

$$Y_t|_{mv} = \begin{bmatrix} 1 - 0.08824q^{-1} & -0.6 - 0.5207q^{-1} \\ 0.5 + 0.07354q^{-1} & 1 + 0.4339q^{-1} \end{bmatrix} a_t$$

This will be the theoretical benchmark to assess performance of the feedback controller. The FCOR algorithm will estimate this benchmark from routine operating data. The results are shown in Figure 8.2. Since y_2 becomes the more important output variable, the benchmark variance of y_2 should be reduced and the benchmark variance of y_1 is expected to increase. Consequently, the performance indices of y_2 should decrease and the performance indices of y_1 is expected to increase. Somehow, the objective function based MIMO performance indices should move toward the performance index of y_2. All these are confirmed by Figure 8.2. This figure also shows good agreement between the theoretical indices and those estimated from the FCOR algorithm. Notice that while individual performance indices can be larger than 1, the objective function based MIMO performance indices are always less than 1.

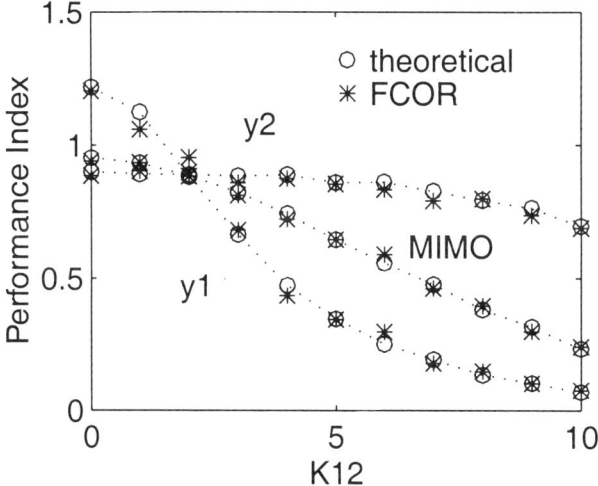

Fig. 8.2. Performance assessment of a square MIMO process (with a general interactor matrix and output weighting) under multiloop minimum variance control

The FCOR algorithm can also handle non-square systems. Consider the input-output transfer function in Example 8.4.1 replaced by a 2×3 transfer function matrix:

$$T = \begin{bmatrix} \frac{q^{-1}}{1-0.4q^{-1}} & \frac{K_{12}q^{-2}}{1-0.1q^{-1}} & \frac{0.2q^{-2}}{1-0.5q^{-1}} \\ \frac{0.7q^{-1}}{1-0.2q^{-1}} & \frac{q^{-2}}{1-0.8q^{-1}} & \frac{0.8q^{-2}}{1-0.7q^{-1}} \end{bmatrix}$$

This non-square transfer function matrix has the following unitary interactor matrix:

$$D = \begin{bmatrix} -0.8192q & -0.5735q \\ -0.5735q^2 & 0.8192q^2 \end{bmatrix}$$

Suppose a multivariable controller:

$$Q = \begin{bmatrix} \frac{0.5-0.20q^{-1}}{1-0.5q^{-1}} & 0 \\ 0 & \frac{0.25-0.200q^{-1}}{(1-0.5q^{-1})(1+0.5q^{-1})} \\ \frac{0.6-0.1q^{-1}}{1-q^{-1}} & \frac{0.6-0.1q^{-1}}{1-q^{-1}} \end{bmatrix}$$

is designed for this system. With the multivariable controller operating on this 2-output and 3-input system, the performance can be estimated by using the FCOR algorithm on routine operating data. The results are shown in Figure 8.3. For this example, the performance index of y_1 steadily deteriorates with increasing interaction while the performance index of y_2 shows a more complicated pattern of performance change with increasing interaction.

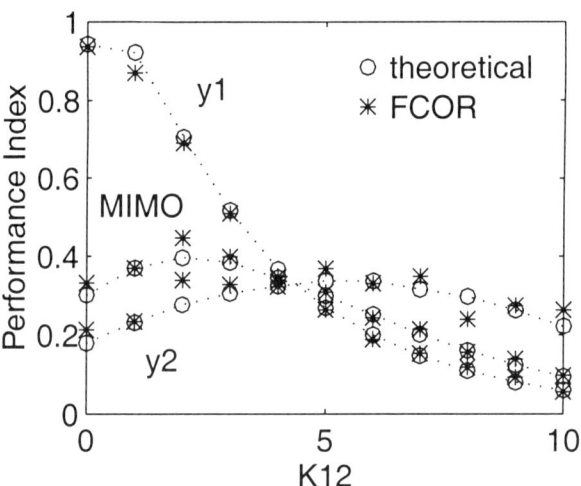

Fig. 8.3. Performance assessment of a non-square MIMO process (with a general interactor matrix) under multivariable control

8.4.2 Industrial application

An industrial absorption process is shown in Figure 8.4. The process is designed for the removal of CO_2 from the feed gas that is a mixture of CO_2, H_2, and N_2. The solution contains a combination of potassium carbonate and a catalyst additive. It absorbs CO_2 from the CO_2 absorber on the right and is regenerated in the CO_2 stripper on the left by reboiling and steam-stripping the CO_2 from the solution. The term "lean solution" refers to stripper bottom

8.4 Evaluation of the FCOR algorithm: simulation and application

flow, which is mostly free of CO_2 whereas the "semi-lean solution" coming from the first-stage stripping still contains some CO_2. Since the solution is circulated continuously between the two towers, an extremely strong interaction exists between the two PID-controlled level loops of both towers. The objective of this analysis is performance assessment of the two level controllers. For this process, the output quality is measured by $J = E(Y_t - Y_t^{sp})^T W (Y_t - Y_t^{sp})$ with the weighting matrix $W = I$ (defined as weighting #1) and the setpoint being constant. Different weighting matrices are also studied in this example. A representative set of routine operating data was sampled over 15 hours with sampling time $T_s = 5s$ as shown in Figure 8.5. The time delay of the loop u_1 -y_1 is known to be 19 sampling periods, and the time delay of the loop u_2 - y_2 is known to be 23 sampling periods. The interactor matrix has the diagonal structure, which is a special unitary interactor matrix. According to plant engineers, the u_1 - y_1 loop is supposed to be tightly tuned. We use this example as an illustrative example to show how the weighting matrix W affects performance assessment of a multivariate process. Experience has shown that a significant number of multivariate processes have or approximately have diagonal interactor matrices.

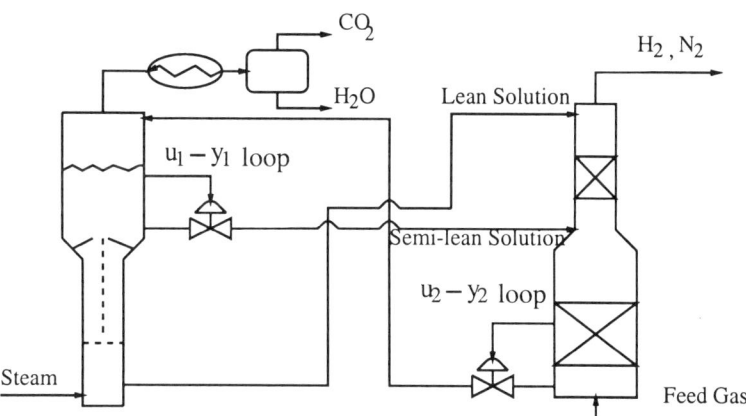

Fig. 8.4. Schematic diagram of the industrial absorption process

The left graph in Figure 8.6 shows the estimated multivariable performance indices (weighting #1 with $W = I$), where each point represents the performance measure based on the last 750 data points, *i.e.*, one hour of data. The performance indices can be seen to be fairly stable over 14 hours with an average multivariable performance index (objective function based) of 0.5. To investigate this process further, the individual output performance is studied. The performance indices of output y_1 and y_2 are also shown in the left graph of Figure 8.6. The performance measure of y_1 is close to minimum variance control(≈ 1). However, the performance measure of y_2 (≈ 0.2) indicates rather poor control of this output. Thus, these two loop tunings are fairly

96   8. Feedback Controller Performance Assessment: General Interactor

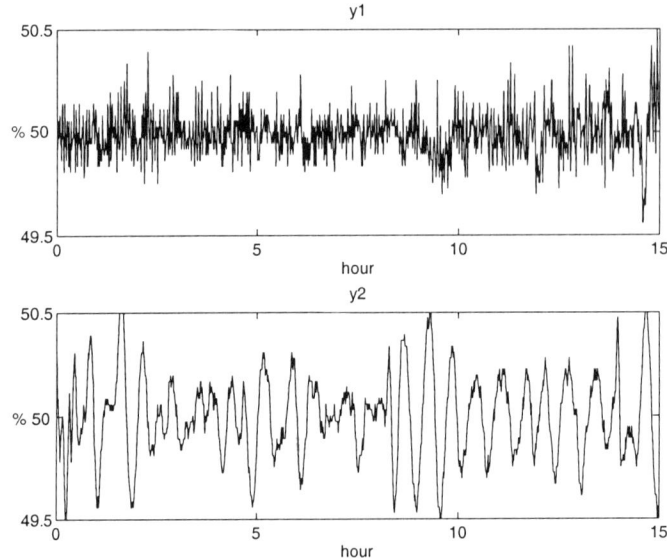

Fig. 8.5. Absorption process data

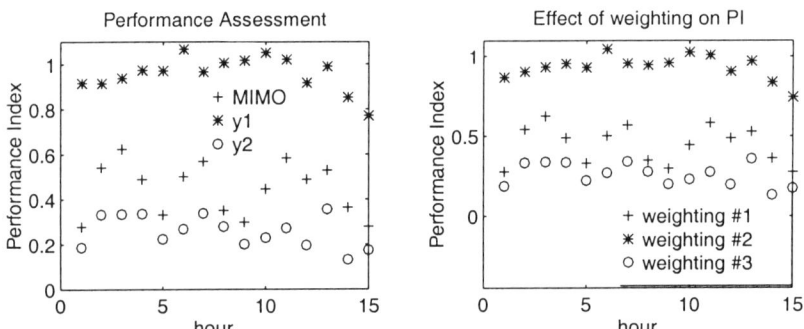

Fig. 8.6. Multivariable performance assessment of absorption process

"unbalanced". Further study shows that a strong negative correlation exists between the variations of y_1 and y_2. This assessment suggests that, therefore: 1) in order to improve the overall performance, the loop 1 controller (for y_1) needs to be detuned so that the loop 2 controller (for y_2) can be more tightly tuned; 2) there may be little or no incentive to improve the performance of y_1 further since it is performing at almost minimum variance levels; and 3) the performance of y_2 may be significantly improved by simply adjusting current control parameters or redesigning the control algorithm, e.g., by detuning the loop 1 controller and tightly tuning the loop 2 controller. In summary, this assessment justifies further analysis of this process, and indicates the potential for improving regulatory performance, particularly the performance of y_2.

Now suppose that the level of the first column is a much more important variable to regulate than the level of the second column. A weighting matrix $W = diag(100, 1)$ (defined as weighting #2) is assumed. One would then expect this process to indicate good performance in terms of objective function-based MIMO performance index. On the other hand, if a weighting matrix $W = diag(1, 100)$ (defined as weighting #3) is assumed, $i.e.$, the level of the second column is much more important than the first column, then one should expect that this process has very poor performance. All these are confirmed in the right side graph in Figure 8.6.

8.5 Summary

The main contributions of this chapter are: 1) the development of a computationally simple algorithm to estimate control loop performance measure of a general class (square and non-square) of MIMO processes; the use of this measure for preliminary process diagnosis and the monitoring of multivariable processes under multiloop control has been illustrated by application to an industrial process; the latter topic is bound to be the subject of considerable industrial interest for pre- and post-audit of advanced control applications; 2) the derivation of this algorithm is based on the idea of a minimum variance benchmark standard that has been extended from the SISO to the MIMO case; 3) the derived algorithm is simple and has been successfully evaluated by simulated and actual industrial application; the proposed performance assessment together with the analysis of dispersion and spectral analysis as introduced by DeVries and Wu (1978) can result in a powerful tool for multivariable performance monitoring and diagnosis.

CHAPTER 9
FEEDFORWARD & FEEDBACK CONTROLLER PERFORMANCE ASSESSMENT

9.1 Introduction

Minimum variance feedback control is the best possible feedback control in the sense that no other controllers can give lower process variance. If a process indicates a good performance measure relative to minimum variance control, further tuning of the existing feedback controller is neither useful nor helpful. If further reduction of process variance is required, then one may have to implement feedforward control.

Design of minimum variance feedforward & feedback control can be found in Box and Jenkins (1976), Sternad and Soderstrom (1988). Desborough and Harris (1993) have discussed feedforward controller performance assessment of SISO processes. The main contribution of this chapter is to extend the MIMO feedback controller performance assessment technique to the feedforward plus feedback case. If feedforward controllers have not actually been implemented on the process, then this analysis gives a measure of the potential benefit of implementing feedforward controllers. This chapter is organized as follows: In Section 9.2, the theoretical background and calculation procedure for FF&FB control performance assessment is established. The proposed method is then illustrated by a numerical example in Section 9.3, followed by concluding remarks in Section 9.4.

9.2 Feedforward & feedback controller performance assessment of MIMO processes

9.2.1 Minimum variance FF&FB control benchmark performance

Consider a MIMO process:

$$Y_t = TU_t + N_a a_t + N_b b_t \tag{9.1}$$

where T $(n \times n)$ is the input-output transfer function matrix, N_a $(n \times n_a)$ and N_b $(n \times n_b)$ are disturbance transfer function matrices, a_t $(n_a \times 1)$ is the "driving force" (white noise) for realization of the unmeasurable disturbances, and b_t ($n_b \times 1$) is the "driving force" for realization of the measurable disturbances which is independent of a_t. The "driving force", b_t, can be measured indirectly through time series analysis (data prewhitening):

$$\xi_t = G_m b_t \tag{9.2}$$

where ξ_t ($n_b \times 1$) is directly measured disturbances, and G_m ($n_b \times n_b$) is the transfer function matrix obtained from time series analysis of ξ_t.

By factoring T as $T = D^{-1}\tilde{T}$ where D^{-1} is the delay matrix or the inverse interactor matrix, Equation 9.1 can be written as

$$Y_t = D^{-1}\tilde{T}U_t + N_a a_t + N_b b_t \tag{9.3}$$

Multiplying both sides of Equation 9.3, by $q^{-d}D$ where d is the order of the interactor matrix (the smallest integer that makes $q^{-d}D$ causal or the largest power of q in D), yields

$$\tilde{Y}_t = q^{-d}\tilde{T}U_t + \tilde{N}_a a_t + \tilde{N}_b b_t \tag{9.4}$$

where

$$\tilde{Y}_t = q^{-d}DY_t$$
$$\tilde{N}_a = q^{-d}DN_a$$
$$\tilde{N}_b = q^{-d}DN_b$$

Using the Diophantine identity

$$\tilde{N}_a = F_a + q^{-d}R_a$$
$$\tilde{N}_b = F_b + q^{-d}R_b$$

where

$$F_a = F_0^{(a)} + F_1^{(a)}q^{-1} + \cdots + F_{d-1}^{(a)}q^{-(d-1)}$$
$$F_b = F_0^{(b)} + F_1^{(b)}q^{-1} + \cdots + F_{d-1}^{(b)}q^{-(d-1)}$$

are matrix polynomials, and R_a and R_b are proper rational transfer function matrices. Using these Diophantine identities, Equation 9.4 can be written as

$$\tilde{Y}_t = F_a a_t + F_b b_t + \tilde{T}U_{t-d} + R_a a_{t-d} + R_b b_{t-d} \tag{9.5}$$

Owing to causality of the control law, U_{t-d} must be independent of $F_a a_t$ and $F_b b_t$, since these two terms occur after the time $t - d$. Therefore, the first two terms on the right hand side of Equation 9.4 are control independent. In other words,

$$Var(\tilde{Y}_t) \geq Var(F_a a_t + F_b b_t)$$

or in LQ form (total variance):

$$E(\tilde{Y}_t^T \tilde{Y}_t) \geq tr(Var(F_a a_t)) + tr(Var(F_b b_t))$$

9.2 FF & FB controller performance assessment: MIMO processes

Equality holds when the remaining terms on the left hand side of Equation 9.5 are equal to zero, *i.e.*,

$$\tilde{T}U_{t-d} + R_a a_{t-d} + R_b b_{t-d} = 0$$

which gives the following control law:

$$U_t = -\tilde{T}^{-1}(R_a a_t + R_b b_t) \tag{9.6}$$

Substituting Equation 9.6 into 9.5 yields

$$\tilde{Y}_t = F_a a_t + F_b b_t$$

This gives

$$a_t = F_a^{-1}(\tilde{Y}_t - F_b b_t) \tag{9.7}$$

Substituting Equation 9.7 into 9.6 results in the feedback plus feedforward control law:

$$U_t = -\tilde{T}^{-1} R_a F_a^{-1}(q^{-d}D)Y_t + \tilde{T}^{-1}(R_a F_a^{-1} F_b - R_b) b_t \tag{9.8}$$

By using Equation 9.2, Equation 9.8 can be written as

$$U_t = -\tilde{T}^{-1} R_a F_a^{-1}(q^{-d}D)Y_t + \tilde{T}^{-1}(R_a F_a^{-1} F_b - R_b) G_m^{-1} \xi_t \tag{9.9}$$

The first term on the right hand side of Equation 9.9 reflects the feedback part of the minimum variance feedforward and feedback control law, and the second term reflects the feedforward part of the minimum variance feedforward and feedback control law. This minimum variance control law may or may not be practical depending on the invertibility of process zeros. In practice, there are many limitations on achievable performance in addition to time-delays and non-minimum phase zeros. Constraints on control action, the existence of poorly damped zeros, and the desired robustness characteristics are examples of such limitations as discussed in Chapter 8. Nevertheless, this minimum variance control provides the absolute lower bound on the process variation and serves as a convenient benchmark for the first-level performance assessment.

Apparently, this minimum variance feedback and feedforward control law only minimizes the total variance (LQ objective function) of the interactor-filtered variable $\tilde{Y}_t$. However, if D is a unitary interactor matrix, then the minimum variance control law, which minimizes the following LQ objective function (total variance) of the interactor-filtered variable $\tilde{Y}_t$:

$$J_1 = E[\tilde{Y}_t^T \tilde{Y}_t] \tag{9.10}$$

also minimizes the LQ objective function of the original variable Y_t:

$$J_2 = E[Y_t^T Y_t] \tag{9.11}$$

and $J_1 = J_2$ (for proof see Lemma 4.3.1. In the sequel, this LQ objective function or total variance will be used for the scalar performance measure of the MIMO system, and we will no longer distinguish the performance measures between the original variable, Y_t, and the interactor-filtered variable, $\tilde{Y}_t$.

In summary, under minimum variance FF & FB control, the closed-loop response can be denoted by

$$e_t \triangleq \tilde{Y}_t|_{min} = F_a a_t + F_b b_t \tag{9.12}$$

where e_t is the FF & FB controller-invariant term, i.e., no feedforward and feedback controllers can change this term, and therefore it can be estimated from routine closed-loop operating data under any linear FF & FB controllers. This yields the following theorem:

Theorem 9.2.1. *The minimum variance of the interactor-filtered output $\tilde{Y}_t$ is FF & FB control invariant, and can be estimated from routine operating data Y_t. The estimation is via multivariate moving-average time series analysis of $\tilde{Y}_t$, i.e., if $\tilde{Y}_t$ is written as*

$$\tilde{Y}_t = \underbrace{F_0^{(a)} a_t + \cdots + F_{d-1}^{(a)} a_{t-d+1}}_{e_t^u = F_a a_t} + F_d^{(a)} a_{t-d} + \cdots +$$

$$+ \underbrace{F_0^{(b)} b_t + \cdots + F_{d-1}^{(b)} b_{t-d+1}}_{e_t^m = F_b b_t} + F_d^{(b)} b_{t-d} + \cdots$$

then

$$e_t = e_t^u + e_t^m$$

constitutes the minimum FF & FB variance, where $e_t^u = F_a a_t$ is the contribution of the unmeasurable disturbances, and $e_t^m = F_b b_t$, the contribution of the measurable disturbances, to the minimum variance sequence e_t.

Note that this minimum FF & FB variance may not be achieved by minimum variance feedback control only, and the lower bound achieved by feedback control is no less than the lower bound achieved by feedforward plus feedback control (Pierce 1975). Estimation of the lower bound achieved by feedback control (minimum FB variance) has been established in Chapter 8, i.e., if one models the interactor-filtered output, $\tilde{Y}_t$, by the following moving-average model:

$$\tilde{Y}_t = \underbrace{F_0 \nu_t + \cdots + F_{d-1} \nu_{t-d+1}}_{e_t^{FB}} + F_d \nu_{t-d} + \cdots$$

then e_t^{FB} is feedback controller-invariant, and $E[(e_t^{FB})^T (e_t^{FB})]$ constitutes the total minimum FB variance. Here ν_t is actually a lumped disturbance from both unmeasurable and measurable disturbances. The benefit of implementing feedforward & feedback control is, therefore,

$$\Delta J = J_{FB} - J_{FF\&FB} = E[(e_t^{FB})^T(e_t^{FB})] - E(e_t^T e_t) \qquad (9.13)$$

or the improvement in performance relative to the present process variance is

$$\Delta J = \frac{E[(e_t^{FB})^T(e_t^{FB})] - E(e_t^T e_t)}{E(Y_t^T Y_t)}$$

or the improvement in performance relative to minimum FB variance is

$$\Delta J = \frac{E[(e_t^{FB})^T(e_t^{FB})] - E(e_t^T e_t)}{E[(e_t^{FB})^T(e_t^{FB})]}$$

9.2.2 Feedback controller performance assessment of MIMO processes using minimum variance FF & FB control as the benchmark

A closed-loop response to both unmeasurable and measurable disturbances can be written as

$$Y_t = G_a a_t + G_\xi \xi_t \qquad (9.14)$$

where G_a and G_ξ are rational, proper transfer function matrices. Equation 9.14 can be estimated via any standard system identification tools with ξ_t as the known input and Y_t as the output. Substituting $\xi_t = G_m b_t$ into Equation 9.14 yields

$$Y_t = G_a a_t + G_\xi G_m b_t \qquad (9.15)$$

where the transfer function matrix G_m can be estimated from multivariate time series analysis of ξ_t via system identification tools. Multiplying Equation 9.15 by $q^{-d}D$ yields

$$q^{-d}DY_t = \tilde{Y}_t = q^{-d}DG_a a_t + q^{-d}DG_\xi G_m b_t \qquad (9.16)$$

Equation 9.16 can be further written in the Markov parameter form (impulse response form):

$$\tilde{Y}_t = \underbrace{F_0^{(a)} a_t + \cdots + F_{d-1}^{(a)} a_{t-(d-1)}}_{e_t^u} + \underbrace{F_d^{(a)} a_{t-d} + \cdots +}_{\hat{e}_t^u}$$

$$+ \underbrace{F_0^{(b)} b_t + \cdots + F_{d-1}^{(b)} b_{t-(d-1)}}_{e_t^m} + \underbrace{F_d^{(b)} b_{t-d} + \cdots}_{\hat{e}_t^m} \qquad (9.17)$$

where $F_i^{(a)}$ and $F_i^{(b)}$ are the Markov parameter matrices (impulse response coefficient matrices); $\hat{e}_t^u$ is the inflation in e_t^u owing to non-optimal FB control to the unmeasurable disturbances; $\hat{e}_t^m$ is the inflation in e_t^m owing to non-optimal FF/FB control of the measurable disturbances. From Equation 9.17, we have

$$Var(e_t^u) = F_0^{(a)}\Sigma_a(F_0^{(a)})^T + \cdots + F_{d-1}^{(a)}\Sigma_a(F_{d-1}^{(a)})^T$$
$$Var(e_t^m) = F_0^{(b)}\Sigma_b(F_0^{(b)})^T + \cdots + F_{d-1}^{(b)}\Sigma_b(F_{d-1}^{(b)})^T$$

where $\Sigma_a = E(a_t a_t^T)$ and $\Sigma_b = E(b_t b_t^T)$. Thus, the quadratic measure (total variance) of minimum FF&FB variance control

$$J_{FF\&FB} = E(e_t^T e_t) = E(e_t^u)^T(e_t^u) + E(e_t^m)^T(e_t^m) \tag{9.18}$$

can be calculated following the above procedure and can then be used as the benchmark against which to assess performance of feedforward & feedback controllers. The procedure for performance assessment of feedforward & feedback control is summarized in Table 9.1. The quadratic measure of minimum FB-only control $J_{FB} = E[(e_t^{FB})^T(e_t^{FB})]$ may be estimated via the FCOR algorithm, after which the benefit of implementing optimal FF & FB control can be calculated from Equation 9.13.

Table 9.1. The procedure for calculation of the minimum variance feedforward & feedback control benchmark performance of MIMO processes

---

1. Treat measured disturbance ξ_t as the input, and fit output data to a model in a form of Equation 9.14.
2. Pre-whitening the measured disturbance using time series analysis in a form of Equation 9.2. Then substitute Equation 9.2 into 9.14 to obtain 9.15.
3. Filter Equation 9.15 by the unitary interactor matrix $q^{-d}D$ to obtain 9.16.
4. Express Equation 9.16 in a Markov parameter form of Equation 9.17.
5. The minimum variance feedforward & feedback control benchmark performance can then be calculated according to Equation 9.18.

9.3 Numerical example

Performance assessment of feedback control for a 2×2 process has been studied in Chapter 8. In this example, the benefit of implementing feedforward and feedback control to the same process will be discussed. The process has the following transfer function matrices:

$$T = \begin{bmatrix} \frac{q^{-1}}{1-0.4q^{-1}} & \frac{q^{-2}}{1-0.1q^{-1}} \\ \frac{0.3q^{-1}}{1-0.1q^{-1}} & \frac{q^{-2}}{1-0.8q^{-1}} \end{bmatrix}$$

$$N_a = \begin{bmatrix} \frac{1}{1-0.5q^{-1}} & \frac{-0.6}{1-0.5q^{-1}} \\ \frac{0.5}{1-0.5q^{-1}} & \frac{1.0}{1-0.5q^{-1}} \end{bmatrix} \quad N_b = \begin{bmatrix} \frac{0.1q^{-2}}{1-0.4q^{-1}} & \frac{-0.2q^{-1}}{1-0.3q^{-1}} \\ \frac{0.3q^{-2}}{1-0.2q^{-1}} & \frac{0.4q^{-1}}{1-0.1q^{-1}} \end{bmatrix}$$

9.3 Numerical example

A unitary interactor matrix D can be factored as

$$D = \begin{bmatrix} -0.9578q & -0.2873q \\ -0.2873q^2 & 0.9578q^2 \end{bmatrix}$$

Then $\tilde{N}_a = q^{-d} D N_a$ is given by

$$\tilde{N}_a = \underbrace{\begin{bmatrix} -1.1014q^{-1} & 0.2874q^{-1} \\ 0.1916 + 0.0958q^{-1} & 1.1302 + 0.5651q^{-1} \end{bmatrix}}_{F_a} +$$

$$q^{-2} \underbrace{\begin{bmatrix} \frac{-0.5507}{1-0.5q^{-1}} & \frac{0.1437}{1-0.5q^{-1}} \\ \frac{0.0479}{1-0.5q^{-1}} & \frac{0.2826}{1-0.5q^{-1}} \end{bmatrix}}_{R_a}$$

and $\tilde{N}_b = q^{-d} D N_b$ is given by

$$\tilde{N}_b = \underbrace{\begin{bmatrix} 0 & 0 \\ 0 & 0.3256q^{-1} \end{bmatrix}}_{F_b} + q^{-2} \underbrace{\begin{bmatrix} \frac{-4.55q^{-1}+1.341q^{-2}}{(5-q^{-1})(5-2q^{-1})} & \frac{-30.65+5.363q^{-1}}{(10-q^{-1})(10-3q^{-1})} \\ \frac{6.465-2.73q^{-1}}{(5-q^{-1})(5-2q^{-1})} & \frac{2.104-0.9768q^{-1}}{(10-q^{-1})(10-3q^{-1})} \end{bmatrix}}_{R_b}$$

The minimum FF & FB variance term can then be written as

$$e_t = F_a a_t + F_b b_t = \begin{bmatrix} -1.1014q^{-1} & 0.2874q^{-1} \\ 0.1916 + 0.0958q^{-1} & 1.1302 + 0.5651q^{-1} \end{bmatrix} a_t +$$

$$\begin{bmatrix} 0 & 0 \\ 0 & 0.3256q^{-1} \end{bmatrix} b_t$$

Assuming $\Sigma_a = E a_t a_t^T = I$ and $\Sigma_b = E b_t b_t^T = 4I$, the quadratic measure (total variance) of minimum variance FF & FB control can be calculated as

$$J_{FF\&FB} = E e_t^T e_t = 3.3626$$

If, on the other hand, only minimum variance feedback control law is implemented, the feedback controller-invariant minimum variance term can be estimated by applying a FB performance assessment algorithm, such as the FCOR algorithm, to process data. This gives the quadratic measure (total variance) of minimum feedback control as

$$J_{FB} = 4.1873$$

The benefit of adding optimal feedforward control relative to minimum FB variance is then

$$\Delta J\% = \frac{4.1873 - 3.3626}{4.1873} = 19.6\%$$

In summary, the total variance achievable by implementing any feedback control is no less than 4.1873; the achievable total variance by implementing any feedback and feedforward control is no less than 3.3626. The benefit of adding feedforward control is about 20% relative to minimum variance FB control. The optimal values may or may not be achievable depending on practical constraints.

9.4 Summary

This chapter has extended the MIMO performance assessment of feedback controllers to feedforward plus feedback controllers. It has been shown that the minimum FF & FB variance is feedforward and feedback controller-invariant and can be estimated from routine operating data via time series analysis. The proposed performance assessment method has been illustrated by a numerical example. The industrial application of this technique can be found in Huang *et al.* (1999).

CHAPTER 10
PERFORMANCE ASSESSMENT OF NONMINIMUM PHASE SYSTEMS

10.1 Introduction

With a complete knowledge of process dynamics, any possible limitations on the achievable performance may be calculated via procedures such as convex optimization and linear programming (Boyd and Barratt 1991; Dahleh and Diaz-Bobillo 1995). This is generally not a very attractive approach to process performance monitoring since a typical plant can have hundreds or thousands of control loops. Identification of all loops, to obtain process models, is a very demanding requirement. Performance monitoring should be carried out in such a way that the normal operation of a process is affected as little as possible. In addition, process dynamics and disturbances may drift from time to time, and the initially identified model may not represent the true dynamics. Thus, on-line performance monitoring is necessary.

Different types of constraints require different levels of process knowledge. Some constraints require less *a priori* knowledge of processes than others. If one can separate the constraints into different levels, then control loop performance may be assessed from the simplest to the hardest constraints with progressively more information required about the process at each stage. Only those loops which indicate poor performance at the first level (screening) test need to be examined at the next level of performance assessment. Time-delays pose the first level of performance limitations. Time-delays are also relatively easy to obtain or estimate. Therefore, performance limitation due to time-delays is assessed at the first level. The second level of performance limitation would be due to non-minimum phase zeros.

Tyler and Morari (1995) have considered performance assessment of SISO systems with non-minimum phase zeros. In Chapters 6, 7 and 8, performance assessment of MIMO processes with time delays or infinite zeros has been considered. This chapter is an extension to the previous results in which we can consider performance limitations and assessment in the presence of unstable process zeros. Throughout this chapter, we shall assume that the process is open-loop stable, *i.e.*, no poles lie outside the unit circle.

This chapter is organized as follows: The generalized unitary interactor matrix, an all-pass factor, is introduced in Section 10.2. Performance assessment of processes with non-minimum phase zeros is discussed in Section 10.3, followed by a numerical example in Section 10.4. Concluding remarks are addressed in Section 10.5.

10.2 Generalized unitary interactor matrices

For multivariable processes with non-minimum phase or non-minimum phase zeros, an interactor matrix which can also factorize the non-minimum phase zeros in addition to the infinite zeros is desirable. The optimal control law corresponding to the admissible minimum variance and minimum ISE control requires such an interactor matrix (Tsiligiannis and Svoronos 1989).

Definition 10.2.1. *An interactor matrix, D_G, satisfying the following four conditions, is defined as the generalized unitary interactor matrix of T.*

1. *The unitary condition is held: $D_G^T(q^{-1})D_G(q) = I$.*
2. *There exists a non-singular constant matrix $K_{inf} \in R^{n \times m}$ such that*

$$\lim_{q^{-1} \to 0} D_G T = K_{inf} \qquad (10.1)$$

3. *There exist non-singular matrices $K_i \in R^{n \times m}$ such that*

$$\lim_{q^{-1} \to 1/\sigma_i} D_G T = K_{f_i} \quad (i = 1, \cdots, s) \qquad (10.2)$$

   where σ_i, $(i = 1, \cdots, s)$ are the non-minimum phase zeros of T, i.e., $|\sigma_i| > 1$.
4. *The poles of D_G in terms of q are $\{\sigma_1, \cdots, \sigma_s\}$ (including the multiplicities).*

The algorithm for factorization of the unitary interactor matrix, as discussed in Chapter 4, extracts infinite zeros from the transfer function matrix. To extract finite non-minimum phase zeros, a bilinear transformation, as introduced by Tsiligiannis and Svoronos (1989) can be used:

$$q^{-1} = \frac{1 + p^{-1}\sigma}{p^{-1} + \sigma}$$

and therefore

$$p^{-1} = \frac{1 - \sigma q^{-1}}{q^{-1} - \sigma}$$

where σ is a non-minimum phase zero, and p^{-1} is a map of q^{-1}. For any non-minimum phase zero $q^{-1} = 1/\sigma$, this mapping transforms the finite zero in the q-domain to an infinite zero in the p-domain, i.e., from $q^{-1} = 1/\sigma$ to $p^{-1} = 0$. Therefore, existing methods for extraction of the infinite zeros can be applied. This also proves the existence of the generalized unitary interactor matrix; therefore, the generalized unitary interactor can be factored as

10.2 Generalized unitary interactor matrices

$$D_G = D_f D_{inf}$$

where D_{inf} is a unitary interactor matrix representing infinite zeros of T, and D_f is a unitary interactor matrix representing non-minimum phase zeros of T. The order (d) of the interactor matrix D_G is defined as the order of D_{inf}. According to this procedure, each p-q transformation factors out one non-minimum phase zero. To factor out all $\{\sigma_1, \cdots, \sigma_s\}$ non-minimum phase zeros, s steps of such transformation are required. Each step involves a simple algebraic manipulation; therefore, D_f can be written as

$$D_f = D_{f_s} D_{f_{s-1}} \cdots D_{f_1}$$

The "Inner-Outer" factorization (Chu 1985) can also factor the non-minimum phase and infinite zeros. It involves the solution of an algebraic Riccati equation in the state space framework. The generalized unitary interactor matrix provides an alternative solution to the inner-outer factorization.

Example 10.2.1. Consider the following system from Tsiligiannis and Svoronos (1989) :

$$T = \begin{bmatrix} \frac{0.6q^{-1}}{1-0.4q^{-1}} & \frac{0.5q^{-1}}{1-0.5q^{-1}} \\ \frac{0.6q^{-1}}{1-0.5q^{-1}} & \frac{0.6q^{-1}}{1-0.4q^{-1}} \end{bmatrix}$$

Tsiligiannis and Svoronos (1989) have shown that a lower triangular generalized interactor matrix can be factored as

$$\begin{bmatrix} q & 0 \\ -1.0954q\frac{1-1.5477q}{q-1.5477} & q\frac{1-1.5477q}{q-1.5477} \end{bmatrix}$$

The optimal control law based on such a lower triangular interactor matrix results in optimal (minimum variance) control of the first variable and conditional optimal control of the remaining variables (Tsiligiannis and Svoronos 1989); therefore, the importance of each variable depends on the order in which it is stacked in the output vector. Different order of the output vector results in the different optimal control laws. On the other hand, as shown in the next section, a generalized unitary interactor matrix gives an optimal control law which minimizes the LQ objective function or H_2 norm, and, therefore, the resulting optimal control law is unique.

Now we show a procedure for factoring a generalized unitary interactor matrix from this process. The transfer function matrix has a simple time-delay structure or a simple interactor matrix $D_{inf} = qI$ (therefore $d = 1$), and a non-minimum phase zero at $1 + \sqrt{0.3} = 1.5477$. If the infinite zeros are factored out, the transfer function matrix with finite (stable and unstable) zeros is obtained as

$$\begin{bmatrix} \frac{0.6q^{-1}}{1-0.4} & \frac{0.5}{1-0.5q^{-1}} \\ \frac{0.6}{1-0.5q^{-1}} & \frac{0.6}{1-0.4q^{-1}} \end{bmatrix}$$

Mapping from q-domain to p-domain using the bilinear transformation

$$q^{-1} = \frac{1 + 1.5477p^{-1}}{1.5477 + p^{-1}}$$

yields the transfer function matrix $(\tilde{T}_f)$ in the p-domain as

$$T = \begin{bmatrix} \frac{0.6(1.5477+p^{-1})}{1.1477+0.3809p^{-1}} & \frac{0.5(1.5477+p^{-1})}{1.0477+0.2262p^{-1}} \\ \frac{0.6(1.5477+p^{-1})}{1.0477+0.2262p^{-1}} & \frac{0.6(1.5477+p^{-1})}{1.1477+0.3809p^{-1}} \end{bmatrix}$$

This can be written in the Markov parameter form as

$$T = \begin{bmatrix} 0.8091 & 0.7386 \\ 0.8863 & 0.8091 \end{bmatrix} + \begin{bmatrix} 0.2542 & 0.3178 \\ 0.3814 & 0.2542 \end{bmatrix} p^{-1} + \cdots$$

The first Markov parameter matrix is rank deficient. Therefore, there is at least one infinite zero in T in the p-domain. Using the first two Markov parameter matrices to form a block Markov parameter matrix yields

$$\Lambda^{(0)} = \begin{bmatrix} 0.8091 & 0.7386 \\ 0.8863 & 0.8091 \\ 0.2542 & 0.3178 \\ 0.3814 & 0.2542 \end{bmatrix}$$

Applying the algorithm given by Equations A.4 to A.7 to $\Lambda^{(0)}$, one can then proceed as follows:

For $i = 1$ (iteration #1): $r_1 = 1, k_1 = 1$,

$$Q^{(1)} = \begin{bmatrix} -0.7385 & 0.6742 \\ -0.6742 & -0.7385 \end{bmatrix}$$

$$U^{(1)} = \begin{bmatrix} 0 & 0 \\ 1 & 0 \\ 0 & 1 \\ 0 & 0 \end{bmatrix}$$

$$U^{(1)}(q) = \begin{bmatrix} 0 & 1 \\ p & 0 \end{bmatrix}$$

$$\Lambda^{(1)} = \begin{bmatrix} -1.2001 & -1.0955 \\ 0.0694 & -0.0633 \\ -0.4531 & -0.4020 \\ 0 & 0 \end{bmatrix}$$

Because $\text{rank}(\Lambda^{(1)}) = 2 = \min(n, m)$, the algorithm terminates and the unitary interactor matrix given by Equations A.2 and A.3 is calculated as

$$D_f(p) = \begin{bmatrix} -0.6742 & -0.7385 \\ -0.7385p & 0.6742p \end{bmatrix}$$

Substituting the bilinear transformation

$$p^{-1} = \frac{1 - 1.5477q^{-1}}{q^{-1} - \sigma}$$

back into $D_f(p)$ yields the unitary interactor matrix in q-domain which contains the non-minimum phase zero of T:

$$D_f = \begin{bmatrix} -0.6742 & -0.7385 \\ \frac{-0.7385(1-1.5477q)}{q-1.5477} & \frac{0.6742(1-1.5477q)}{q-1.5477} \end{bmatrix}$$

Therefore a generalized unitary interactor matrix, containing both infinite and finite non-minimum phase zeros, can be calculated as

$$D_G = D_f D_{inf} = \begin{bmatrix} -0.6742q & -0.7385q \\ \frac{-0.7385(1-1.5477q)q}{q-1.5477} & \frac{0.6742(1-1.5477q)q}{q-1.5477} \end{bmatrix}$$

10.3 Performance assessment of MIMO non-minimum phase processes

10.3.1 Performance assessment with admissible minimum variance control as the benchmark

For MIMO systems with non-minimum phase zeros, the LQ problem can be solved via spectral factorization (Youla and Bongiorno 1985; Harris and MacGregor 1987; Peng and Kinnaert 1992), via optimal H_2 control (Morari and Zafiriou 1989; Dahleh and Diaz-Bobillo 1995) or via the state space approach (Kwakernaak and Sivan 1972). Alternatively, one may solve it through a simple and intuitive approach as discussed in Astrom and Wittenmark (1990). This last approach provides an explicit expression for the feedback controller invariant terms, and is the most suitable approach for seeking the feedback control invariant term (benchmark control performance) for control loop performance assessment. Astrom and Wittenmark (1990) have shown that this admissible minimum variance control problem for SISO systems can be solved by minimizing the filtered variable y_t^f. The filter is an all-pass factor, which removes all zeros that are outside the unit circle from the input-output transfer function while keeping the spectrum unchanged; therefore, the spectrum of y_t^f is the same as that of y_t. Minimization of $Var(y_t^f)$ is equivalent to minimization of $Var(y_t)$. The generalized unitary interactor matrix is also an all-pass factor and can serve as such a filter for the MIMO system as well. Thus, the methodology used in the SISO case can be extended to the MIMO case in an intuitive way. The following theorem is an extension to Astrom and Wittenmark(1990) and to Goodwin and Sin(1984) by considering the generalized unitary interactor matrix for the solution of the admissible minimum variance control law. The admissible minimum variance control law can also be derived from the optimal H_2 control law. This is discussed in Section 10.3.2.

10. Performance Assessment of Nonminimum Phase Systems

Theorem 10.3.1. *Consider a MIMO process with non-minimum phase zeros*

$$Y_t = TU_t + Na_t \tag{10.3}$$

The control objective is to minimize the LQ objective function defined by

$$J = E(Y_t^T Y_t) \tag{10.4}$$

Then the admissible minimum variance control law is given by

$$U_t = -\tilde{T}^{-1} R_{mp} (F + q^{-d} R_{nmp})^{-1} (q^{-d} D_G) Y_t \tag{10.5}$$

where $\tilde{T} = D_G T$; d is the order of the interactor matrix; D_G is the generalized unitary interactor; F, R_{mp} and R_{nmp} are derived from the Diophantine identity:

$$q^{-d} D_G N = \underbrace{F_0 + F_1 q^{-1} + \cdots + F_{d-1} q^{-d+1}}_{F} + q^{-d} \underbrace{(R_{nmp} + R_{mp})}_{R} \tag{10.6}$$

where F_i (for $i = 0, 1, \cdots, d-1$) are constant coefficient matrices, R is the remaining proper transfer matrix, R_{nmp} contains all unstable poles of R after partial fraction expansion, and R_{mp} is the remaining term of R after the partial fraction expansion.

Proof. Multiplying both sides of Equation 10.3 by $q^{-d} D_G$ yields

$$q^{-d} D_G Y_t = \tilde{T} U_{t-d} + q^{-d} D_G N a_t \tag{10.7}$$

Substituting Equation 10.6 into 10.7 yields

$$q^{-d} D_G Y_t = \tilde{T} U_{t-d} + F a_t + q^{-d} R_{nmp} a_t + q^{-d} R_{mp} a_t \tag{10.8}$$

Since $G_G^T(q^{-1}) G_G(q) = I$ by the definition of the generalized unitary interactor, the minimization of $E[(q^{-d} D_G Y_t)^T (q^{-d} D_G Y_t)]$ is equivalent to the minimization of $E[Y_t^T Y_t]$ for any admissible feedback control.

The following interpretation of the unstable operator follows from Astrom and Wittenmark (1990) (also see Wiener(1949)). Consider the operator $1/(1 + aq^{-1})$ where $|a| > 1$. This operator is normally interpreted as a causal unstable operator. Because $|a| > 1$ and the shift operator has the norm $|q| = 1$, the series expansion

$$\frac{1}{1 + aq^{-1}} = \frac{1}{a} \frac{q}{1 + q/a} = \frac{q}{a}[1 - \frac{1}{a}q + \frac{1}{a^2}q^2 - \cdots]$$

converges. Thus, the operator $1/(1 + aq^{-1})$ can be interpreted as a non-causal stable operator; therefore, the term, $q^{-d} R_{nmp}$, in Equation 10.8 can be expanded in terms of the 'q' operator. For example, consider $R_{nmp} = 1/(1 + aq^{-1})$. Then the expansion

10.3 Performance assessment of MIMO non-minimum phase processes

$$\frac{q^{-d}}{1+aq^{-1}} = \frac{1}{a}\frac{q^{-d+1}}{1+q/a} = \frac{1}{a}[q^{-d+1} - \frac{1}{a}q^{-d+2} + \frac{1}{a^2}q^{-d+3} - \cdots]$$

converges, and the most recent term in this expansion is $\frac{1}{a}q^{-d+1}$ that cannot be controlled by the control action, $U_{t-d} = U_t q^{-d}$, which is one step earlier. Clearly the term Fa_t is also independent of U_{t-d}; therefore, the optimal control law is obtained by setting the sum of the remaining terms in Equation 10.8 to zero. This yields

$$U_{t-d} = -\tilde{T}^{-1} R_{mp} a_{t-d} \tag{10.9}$$

or

$$U_t = -\tilde{T}^{-1} R_{mp} a_t \tag{10.10}$$

Substituting Equation 10.10 into Equation 10.8 yields

$$q^{-d} D_G Y_t = (F + q^{-d} R_{nmp}) a_t \tag{10.11}$$

Thus,

$$a_t = (F + q^{-d} R_{nmp})^{-1} (q^{-d} D_G) Y_t \tag{10.12}$$

Substituting Equation 10.12 into 10.10 yields the admissible minimum variance control law:

$$U_t = -\tilde{T}^{-1} R_{mp} (F + q^{-d} R_{nmp})^{-1} (q^{-d} D_G) Y_t \tag{10.13}$$

From Equation 10.11, the closed-loop response under the admissible minimum variance control is given by

$$Y_t = (q^{-d} D_G)^{-1} (F + q^{-d} R_{nmp}) a_t \triangleq G_{min} a_t \tag{10.14}$$

Note that the non-minimum phase zeros of the process are retained and not canceled under closed-loop control. Now we are in a position to show an approach to estimate the admissible minimum variance control variance from closed-loop data using the results in Theorem 10.3.1. It is clear from the proof of Theorem 10.3.1 that, for such a purpose, one needs to estimate the terms F and R_{nmp} from closed-loop data in order to obtain Equation 10.11 or Equation 10.14.

If D_G contains only infinite zeros, then Equation 10.11 reduces to

$$q^{-d} D_G Y_t = F a_t$$

Any non-optimal feedback control will inflate the process by adding an extra term to this equation as discussed in Chapter 8, *i.e.*,

$$q^{-d} D_G Y_t = F a_t + L a_{t-d}$$

where L is a proper rational transfer function matrix and is feedback control dependent. The correlation analysis between the interactor-filtered variable

$q^{-d}D_G Y_t$ and a_t yields the feedback controller-invariant term F. Thus a simple FCOR-based correlation analysis yields the benchmark performance. If D_G contains unstable poles (non-minimum phase zeros of T), however, Equation 10.11 is the closed-loop response under admissible minimum variance control. It is evident from the proof of Theorem 10.3.1 that the terms F and R_{nmp} are feedback control invariant, and any non-optimal feedback control will add an extra term to Equation 10.11 as

$$q^{-d}D_G Y_t = Fa_t + R_{nmp}a_{t-d} + La_{t-d}$$

where L is feedback control dependent.

To estimate the terms F and R_{nmp}, we shall fit Y_t by a time series model as

$$Y_t = G_{cl}a_t$$

Then multiply G_{cl} by $q^{-d}D_G$ to obtain the interactor-filtered closed-loop transfer function matrix G'_{cl}, i.e.,

$$G'_{cl} = q^{-d}D_G G_{cl}$$

Let G'_{cl} be expanded to

$$G'_{cl} = Fa_t + \Phi a_{t-d}$$

where $F = F_0 + F_1 q^{-1} + \cdots + F_{d-1}q^{-d+1}$, and Φ is the remaining rational proper transfer function matrix of G'_{cl}. Finally, from Φ one can obtain R_{nmp} as

$$R_{nmp} = \{\Phi\}_+$$

where $\{.\}_+$ denotes that after a partial fraction expansion of the operand, only the terms corresponding to unstable poles are retained. With the knowledge of F and R_{nmp}, the closed-loop response under the admissible minimum variance control can be calculated from Equation 10.14.

The algorithm to calculate the admissible minimum variance control response for feedback control performance assessment of nonminimum phase processes is summarized in Table 10.1

10.3.2 Alternative proof of admissible minimum variance control

To justify the proof and also the interpretation of causal unstable operators adopted in Theorem 10.3.1, we compare the control law obtained in Theorem 10.3.1 with the optimal H_2 control law. Morari and Zafiriou (1989) have solved minimum H_2-norm control for the MIMO system with non-minimum phase zeros. In this section, we show that the admissible minimum variance control law given in Theorem 10.3.1 is the same as the optimal H_2 control law given by Morari and Zafiriou (1989) for stochastic systems.

10.3 Performance assessment of MIMO non-minimum phase processes

Theorem 10.3.2. *(Morari and Zafiriou 1989): Consider the MIMO process with non-minimum phase zeros,*

$$Y_t = TU_t + Na_t$$

Factor T into all-pass portion and minimum-phase portion

$$T = D_G^{-1}\tilde{T}$$

where D_G^{-1} or D_G is an all-pass factor. Similarly, factor N into

$$N = \tilde{N}N_p$$

where N_p is an all-pass factor. Then, the H_2 optimal control is given by

$$Q^* = q\tilde{T}^{-1}\{q^{-1}D_G\tilde{N}\}_*\tilde{N}^{-1}$$

where Q^* is the controller transfer function matrix in the IMC framework. Its relation with the conventional feedback control Q $(U_t = -QY_t)$ is given by

$$Q = Q^*(I - TQ^*)^{-1}$$

The operator $\{.\}_*$ denotes that after a partial fraction expansion of the operand and with the exception of those corresponding to the poles of D_G, only the strictly proper terms are retained.

Proof. See Morari and Zafiriou (1989).

Assume N has no zeros outside the unit circle. This is a general assumption for the stochastic system (Astrom and Wittenmark 1990; Goodwin and Sin 1984) since any non-minimum phase zeros in N can be replaced by their reciprocals without changing the disturbance spectrum. With this assumption, the control law under a conventional feedback control framework can be written as

$$\begin{aligned}
Q &= Q^*(I - TQ^*)^{-1} & (10.15) \\
&= q\tilde{T}^{-1}\{q^{-1}D_GN\}_*N^{-1}(I - qT\tilde{T}^{-1}\{q^{-1}D_GN\}_*N^{-1})^{-1} \\
&= q\tilde{T}^{-1}\{q^{-1}D_GN\}_*(N - qD_G^{-1}\{q^{-1}D_GN\}_*)^{-1} \\
&= q\tilde{T}^{-1}\{q^{-1}D_GN\}_*(D_GN - q\{q^{-1}D_GN\}_*)^{-1}D_G & (10.16)
\end{aligned}$$

With use of the Diophantine identity:

$$q^{-d}D_GN = \underbrace{F_0 + F_1q^{-1} + \cdots + F_{d-1}q^{-d+1}}_{F} + q^{-d}\underbrace{(R_{nmp} + R_{mp})}_{R}$$

we have

$$\begin{aligned}
q^{-1}D_GN &= q^{d-1}(q^{-d}D_GN) \\
&= q^{d-1}F + q^{-1}R_{nmp} + q^{-1}R_{mp} & (10.17)
\end{aligned}$$

In Equation 10.17, only the term $q^{-1}R_{mp}$ is strictly proper without containing poles of D_G. Note that poles of D_G are the non-minimum phase zeros of T. Therefore,

$$\{q^{-1}D_G N\}_* = q^{-1}R_{mp} \tag{10.18}$$

and consequently

$$D_G N - q\{q^{-1}D_G N\}_* = q^d F + R_{nmp} \tag{10.19}$$

Substituting Equations 10.18 and 10.19 into Equation 10.16 yields

$$\begin{aligned} Q &= q\tilde{T}^{-1}q^{-1}R_{mp}(q^d F + R_{nmp})^{-1}D_G \\ &= \tilde{T}^{-1}R_{mp}(F + q^{-d}R_{nmp})^{-1}(q^{-d}D_G) \end{aligned}$$

Thus, the optimal H_2 control law expressed in the conventional feedback control framework can be written as

$$U_t = -\tilde{T}^{-1}R_{mp}(F + q^{-d}R_{nmp})^{-1}(q^{-d}D_G)Y_t$$

which is the same as the admissible minimum variance control law in Equation 10.13.

Table 10.1. The procedure for calculation of the benchmark performance of MIMO processes with non-minimum phase zeros

---

1. estimate or factorize the generalized unitary interactor matrix from T as D_G, and determine d;
2. fit routine operating data Y_t by a time series model to obtain G_{cl};
3. multiply G_{cl} by $q^{-d}D_G$ to obtain $G'_{cl} = q^{-d}D_G G_{cl}$, where d is the order of the interactor matrix;
4. expand G'_{cl} into

$$G'_{cl} = \underbrace{F_0 + F_1 q^{-1} + \cdots + F_{d-1}q^{-(d-1)}}_{F} + q^{-d}\phi$$

where F_i for $(i = 1, 2, \cdots, d-1)$ are constant coefficient matrices, and ϕ is the remaining term after the expansion;

5. using partial fraction expansion, ϕ can be expanded into

$$\phi = R_{nmp} + L$$

where R_{nmp} contains all unstable poles which are the non-minimum phase zeros of T, and L is the remaining term after the partial fraction expansion. Then the process under admissible minimum variance control can be written as

$$Y_t = q^d D_G^{-1}(F + q^{-d}R_{nmp})a_t \tag{10.20}$$

10.4 Numerical example

Example 10.4.1. Consider the same system as in Example 10.2.1 with the disturbance transfer function matrix as

$$N = \begin{bmatrix} \frac{1}{1-0.5q^{-1}} & \frac{-0.6}{1-0.5q^{-1}} \\ \frac{0.5}{1-0.5q^{-1}} & \frac{1}{1-0.5q^{-1}} \end{bmatrix}$$

A generalized unitary interactor matrix has been factored in Example 10.2.1 as

$$D_G = D_f D_{inf} = \begin{bmatrix} -0.6742q & -0.7385q \\ \frac{-0.7385(1-1.5477q)q}{q-1.5477} & \frac{0.6742(1-1.5477q)q}{q-1.5477} \end{bmatrix}$$

With the order of the interactor matrix $d = 1$, we have the Diophantine identity

$$q^{-d} D_G N = q^{-1} D_G N$$

$$= \underbrace{\begin{bmatrix} -1.0434 & -0.3340 \\ 0.6213 & -1.7293 \end{bmatrix}}_{F} +$$

$$+ q^{-1} \underbrace{\begin{bmatrix} \frac{-0.5217}{1-0.5q^{-1}} & \frac{-0.167}{1-0.5q^{-1}} \\ \frac{0.8708-0.4808q^{-1}}{(1-0.5q^{-1})(1-1.5477q^{-1})} & \frac{-2.4238+1.3383q^{-1}}{(1-0.5q^{-1})(1-1.5477q^{-1})} \end{bmatrix}}_{R}$$

$$= \underbrace{\begin{bmatrix} -1.0434 & -0.3340 \\ 0.6213 & -1.7293 \end{bmatrix}}_{F} + q^{-1} \underbrace{\begin{bmatrix} 0 & 0 \\ \frac{0.8275}{1-1.5477q^{-1}} & \frac{-2.3032}{1-1.5477q^{-1}} \end{bmatrix}}_{R_{nmp}} +$$

$$+ q^{-1} \underbrace{\begin{bmatrix} \frac{-0.5217}{1-0.5q^{-1}} & \frac{-0.167}{1-0.5q^{-1}} \\ \frac{0.0433}{1-0.5q^{-1}} & \frac{-0.1206}{1-0.5q^{-1}} \end{bmatrix}}_{R_{mp}}$$

$$= \underbrace{\begin{bmatrix} -1.0434 & -0.3340 \\ \frac{0.6213-0.1340q^{-1}}{1-1.5477q^{-1}} & \frac{-1.7293+0.3731q^{-1}}{1-1.5477q^{-1}} \end{bmatrix}}_{F+q^{-d}R_{nmp}} + q^{-1} \underbrace{\begin{bmatrix} \frac{-0.5217}{1-0.5q^{-1}} & \frac{-0.1670}{1-0.5q^{-1}} \\ \frac{0.0433}{1-0.5q^{-1}} & \frac{-0.1206}{1-0.5q^{-1}} \end{bmatrix}}_{R_{mp}}$$

(10.21)

The closed-loop response under admissible minimum variance control is therefore

$$q^{-d} D_G Y_t = (F + q^{-d} R_{nmp}) a_t$$

Substituting numerical values, we have

118   10. Performance Assessment of Nonminimum Phase Systems

$$\begin{bmatrix} -0.6742 & -0.7385 \\ \frac{-0.7385(q^{-1}-1.5477)}{1-1.5477q^{-1}} & \frac{0.6472(q^{-1}-1.5477)}{1-1.5477q^{-1}} \end{bmatrix} Y_t$$

$$= \begin{bmatrix} -1.0434 & -0.3340 \\ \frac{0.6213-0.1340q^{-1}}{1-1.5477q^{-1}} & \frac{-1.7293+0.3731q^{-1}}{1-1.5477q^{-1}} \end{bmatrix} a_t$$

This can be simplified as

$$\begin{bmatrix} -0.6742 & -0.7385 \\ -0.7385(q^{-1}-1.5477) & 0.6472(q^{-1}-1.5477) \end{bmatrix} Y_t$$

$$= \begin{bmatrix} -1.0434 & -0.3340 \\ 0.6213-0.1340q^{-1} & -1.7293+0.3731q^{-1} \end{bmatrix} a_t \quad (10.22)$$

Equation 10.22 represents the theoretical closed-loop response under admissible minimum variance control. Assume, for simplicity, that $Var(a_t) = I$, then the achievable minimum variance can be calculated from Equation 10.22 as

$$Var(Y_t)|_{mv} = \begin{bmatrix} 1.6137 & -0.3521 \\ -0.3521 & 1.4988 \end{bmatrix} \triangleq \Sigma_{ach} \quad (10.23)$$

Define the individual output variance under the admissible minimum variance control as

$$[\sigma_{1,ach}^2, \sigma_{2,ach}^2] \triangleq diag\{\Sigma_{ach}\} \quad (10.24)$$

Now consider the calculation of this achievable minimum variance from the closed-loop transfer function under feedback control. An IMC controller (Tsiligiannis and Svoronos 1989) given by

$$Q^* = \frac{(1-0.4q^{-1})(1-0.5q^{-1})}{(1-0.45q^{-1})(q^{-1}-1.55)} \begin{bmatrix} 7.78(1-0.36q^{-1}) & -8.33(1-0.4q^{-1}) \\ -12.43(1-0.52q^{-1}) & 10(1-0.5q^{-1}) \end{bmatrix}$$

is implemented on the process. The controller transfer function matrix Q^* denotes the control under IMC framework. Under IMC control, the closed-loop transfer function, which can be estimated via time series analysis in practice, is written as

$$G_{cl} = (I - TQ^*)N$$

$$= \begin{bmatrix} 1-q^{-1} & 0 \\ \frac{-2.7908(1-q^{-1})q^{-1}}{1.5477-q^{-1}} & \frac{1.5477(1-q^{-2})}{1.5477-q^{-1}} \end{bmatrix} \begin{bmatrix} \frac{1}{1-0.5q^{-1}} & \frac{-0.6}{1-0.5q^{-1}} \\ \frac{0.5}{1-0.5q^{-1}} & \frac{1}{1-0.5q^{-1}} \end{bmatrix}$$

$$(10.25)$$

The interactor-filtered closed-loop transfer function matrix can be written as

$$G'_{cl} = q^{-d}D_G G_{cl} = q^{-1}D_G G_{cl}$$

$$= \begin{bmatrix} \frac{-1.6150+3.7787q^{-1}-2.1637q^{-2}}{(1-0.5q^{-1})(1.5477-q^{-1})} & \frac{-0.5169-2.2672q^{-1}+2.7841q^{-2}}{(1-0.5q^{-1})(1.5477-q^{-1})} \\ \frac{0.6213-0.6213q^{-2}}{(1-0.5q^{-1})(1-1.5477q^{-1})} & \frac{-1.7293+1.7293q^{-2}}{(1-0.5q^{-1})(1-1.5477q^{-1})} \end{bmatrix}$$

This can be expressed via the Diophantine identity as

$$G'_{cl} = \underbrace{\begin{bmatrix} -1.0434 & -0.3340 \\ 0.6213 & -1.7293 \end{bmatrix}}_{F} +$$

$$+ q^{-1} \underbrace{\begin{bmatrix} \frac{1.9277-1.6420q^{-1}}{(1-0.5q^{-1})(1.5477-q^{-1})} & \frac{-2.8596+2.9511q^{-1}}{(1-0.5q^{-1})(1.5477-q^{-1})} \\ \frac{1.2722-1.1021q^{-1}}{(1-0.5q^{-1})(1-1.5477q^{-1})} & \frac{-3.5411+3.0676q^{-1}}{(1-0.5q^{-1})(1-1.5477q^{-1})} \end{bmatrix}}_{R}$$

$$= \underbrace{\begin{bmatrix} -1.0434 & -0.3340 \\ 0.6213 & -1.7293 \end{bmatrix}}_{F} + q^{-1} \underbrace{\begin{bmatrix} 0 & 0 \\ \frac{0.8275}{1-1.5477q^{-1}} & \frac{-2.3032}{1-1.5477q^{-1}} \end{bmatrix}}_{R_{nmp}} +$$

$$+ q^{-1} \underbrace{\begin{bmatrix} \frac{-0.5217}{1-0.5q^{-1}} & \frac{-0.167}{1-0.5q^{-1}} \\ \frac{0.0433}{1-0.5q^{-1}} & \frac{-0.1206}{1-0.5q^{-1}} \end{bmatrix}}_{L}$$

$$= \underbrace{\begin{bmatrix} -1.0434 & -0.3340 \\ \frac{0.6213-0.1340q^{-1}}{1-1.5477q^{-1}} & \frac{-1.7293+0.3731q^{-1}}{1-1.5477q^{-1}} \end{bmatrix}}_{F+q^{-d}R_{nmp}} +$$

$$+ q^{-1} \underbrace{\begin{bmatrix} \frac{1.9277-1.6420q^{-1}}{(1-0.5q^{-1})(1.5477-q^{-1})} & \frac{-2.8596+2.9511q^{-1}}{(1-0.5q^{-1})(1.5477-q^{-1})} \\ \frac{0.4448}{1-0.5q^{-1}} & \frac{-1.2380}{1-0.5q^{-1}} \end{bmatrix}}_{L}$$

The first term on the left hand side of the last equation is the same as that given in Equation 10.21; therefore, one can see that, indeed, the minimum achievable variance term $F + q^{-d}R_{nmp}$ can be estimated from closed-loop data. In practice, the estimation of this term requires time series analysis of closed-loop data Y_t and the *a priori* knowledge of the generalized unitary interactor matrix D_G.

Assuming that $Var(a_t) = I$, the closed-loop output variance can be calculated from Equation 10.25 as

$$Var(Y_t) = \begin{bmatrix} 1.8133 & -1.1546 \\ -1.1546 & 9.5164 \end{bmatrix} \triangleq \Sigma_Y \qquad (10.26)$$

Comparing Equation 10.26 with Equation 10.23 allows one to compare the actual variance with the achievable minimum variance. The objective function based performance index as defined in Chapter 8 can be calculated as

$$\eta \triangleq \frac{min(E[Y_t^T Y_t])}{E[Y_t^T Y_t]} = \frac{tr\Sigma_{ach}}{tr\Sigma_Y} = \frac{1.6137 + 1.4988}{1.8133 + 9.5164} = 0.27$$

which indicates an overall MIMO feedback control performance. With the maximum performance index as 1 and poorest performance index as 0, this index indicates relatively poor performance. The reason is that, according to Tsiligiannis and Svoronos (1989), the controller is actually designed for setpoint tracking of a step change but not for regulating the disturbances as assumed in this example. Nevertheless, output #1 is close to its lower bound with the individual performance index as

$$\eta_1 \triangleq \frac{\sigma^2_{1,ach}}{\sigma^2_{y_1}} = \frac{1.6137}{1.8133} = 0.89$$

where $\sigma^2_{y_1}$ is the individual output variance defined by

$$[\sigma^2_{y_1}, \sigma^2_{y_2}] \triangleq diag(\Sigma_Y)$$

Output #2, on the other hand, is far away from its lower bound with the individual performance index as

$$\eta_2 \triangleq \frac{\sigma^2_{2,ach}}{\sigma^2_{y_2}} = \frac{1.4988}{9.5164} = 0.16$$

Thus, the controller has good performance in regulating output #1 but poor performance in regulating output #2. The reason is, as shown in Tsiligiannis and Svoronos (1989), that a lower triangular interactor matrix was used for the control design and, therefore, good performance of the first output would be expected.

10.5 Summary

A generalized unitary interactor matrix has been introduced in this chapter. The admissible minimum variance control law derived by using the generalized unitary interactor matrix has been shown to be identical to the optimal H_2 control law. With *a priori* knowledge of the generalized unitary interactor matrix, the admissible minimum variance control performance can be estimated from routine operating data, and used subsequently for control loop performance assessment. A numerical example demonstrates the applicability of the proposed method.

CHAPTER 11
A UNIFIED APPROACH TO PERFORMANCE ASSESSMENT

11.1 Introduction

Feedback control performance assessment with minimum variance control as the benchmark has been discussed in the earlier chapters. It has been shown that this technique is an efficient and also the most convenient tool to monitor industrial processes which can have hundreds and even thousands of control loops.

If the object is not stochastic control but, for example, step disturbance rejection or setpoint tracking, it has been shown that the technique cited above gives an inadequate measure of performance(Eriksson and Isaksson 1994) Tyler and Morari (1995) make a similar claim on this issue. One objective of this chapter is to discuss this issue and extend Harris' idea of control loop performance assessment to cover practical issues such as deterministic disturbances and setpoint changes. It will be shown that many practical problems such as those posed by Eriksson and Isaksson and others can be solved readily under the framework proposed by Harris (1989) via an appropriate formulation of the initial problem. Another objective of this chapter is to unify the performance assessment of both stochastic and deterministic systems under the H_2 norm framework; therefore, the results developed in the earlier chapters can be extended to more general cases.

This chapter is organized as follows: Assessment of setpoint tracking performance is discussed in Section 11.2. In Section 11.3, deterministic disturbances are explained under the stochastic framework. Performance assessment of feedback controllers for regulating both stochastic and deterministic disturbances is discussed in Section 11.4, and the treatment on pure deterministic disturbances is discussed in Section 11.5. In Section 11.6, a unified approach for performance assessment is proposed. A simulated example is given in Section 11.7, followed by concluding remarks in Section 11.8.

11.2 Setpoint tracking problem

As discussed in the previous chapters, the standard formulation of performance assessment using minimum variance control as the benchmark is shown in Figure 11.1 by assuming $Y_t^{sp} = 0$ or $\xi_t = 0$, where N, Q, T and M are disturbance, controller, plant and setpoint transfer function matrices respectively; and a_t is white noise with zero mean; Using this assumption it follows from Figure 11.1 that

$$Y_t = -TQY_t + Na_t \tag{11.1}$$

Under this formulation, routine operating data Y_t can be used to performance assessment. However, this formulation is of interest only when applied for performance assessment of the regulatory controller. In some cases, the setpoint tracking performance may also be of interest.

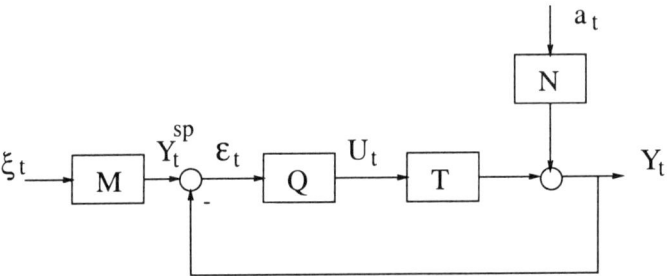

Fig. 11.1. Block diagram of closed-loop process

Define $\epsilon_t = Y_t^{sp} - Y_t$ as the setpoint tracking error. Assuming $a_t = 0$ for consideration of a pure setpoint tracking problem[1], it follows from Figure 11.1 that

$$\epsilon_t = -TQ\epsilon_t + M\xi_t \tag{11.2}$$

where the setpoint, Y_t^{sp}, can be regarded as the realization of a white noise sequence, ξ_t, which is an input of a rational transfer function matrix, M. It will be shown that a "deterministic" setpoint can also be produced by filtering a white noise input. One may note that Equation 11.2 has the same form as Equation 11.1, i.e., the setpoint tracking problem can be formulated as a regulatory control problem by a simple change of variables. The only difference is the data used for analysis. For the setpoint tracking problem, one uses $\epsilon_t = Y_t^{sp} - Y_t$, while either ϵ_t or Y_t can be used for the performance assessment of the regulatory control problem. In the sequel, we will focus on the regulatory control problem, i.e., we will consider performance assessment of Y_t with $Y_t^{sp} = 0$. If setpoint tracking is also considered, then Y_t simply needs to be replaced by the tracking error $\epsilon_t = Y_t^{sp} - Y_t$, and the results developed in the previous chapters can be used.

---

[1] If $a_t \neq 0$, the setpoint signal needs to be considered with other disturbances. This issue will be discussed in the following sections.

11.3 Deterministic disturbances occurring at random time

MacGregor et al. (1984) have shown that many deterministic disturbances such as the step, ramp and exponential changes can be modeled as autoregressive integrated moving average (ARIMA) processes. The only difference between deterministic and stochastic disturbances is the probability distribution of the *shocks* or the white-noise sequence. This point is illustrated by using the following example:

Three ARIMA filters are studied in this example, *i.e.*, the integral filter $1/(1-q^{-1})$, the double integral filter $1/(1-q^{-1})^2$ and the sinusoidal filter $sin(\omega)q^{-1}/(1-2q^{-1}cos(\omega)+q^{-2})$. Figure 11.2 shows probability density function of the shock in a special case. It is symmetric but not normally distributed. The shock takes only three values, -1, 0 and 1 with 99.98% probability density concentrated at the point 0. Figure 11.3 shows the realization of the signal by passing the shock through the three filters. Clearly, these graphs represent a deterministic step, ramp, or sinusoidal signal with possible magnitude, slope, or phase changes respectively occurring at random times. The magnitude change of the sinusoidal signal in Figure 11.3 is due to the non-zero initial value when the second shock occurs. The true deterministic signal (occurring at random time) is obtained in the limiting case. Readers can also refer to (Ljung 1998) for discussion of such deterministic signals for identification problems.

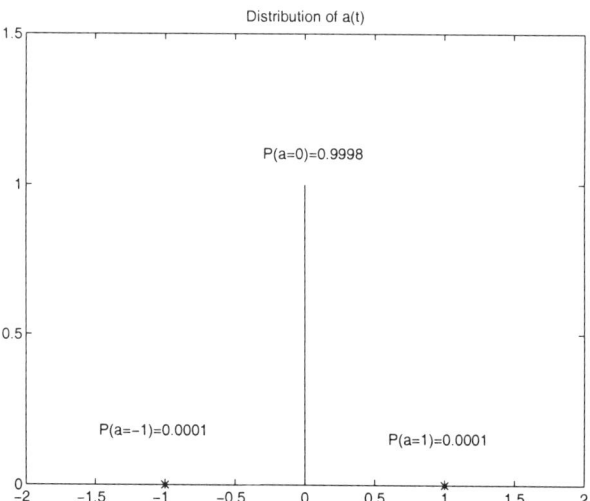

Fig. 11.2. Probability distribution of the shock

Optimal stochastic control laws, such as MVC, GPC and LQG, are independent of the probability distribution of the shock as long as the shock has

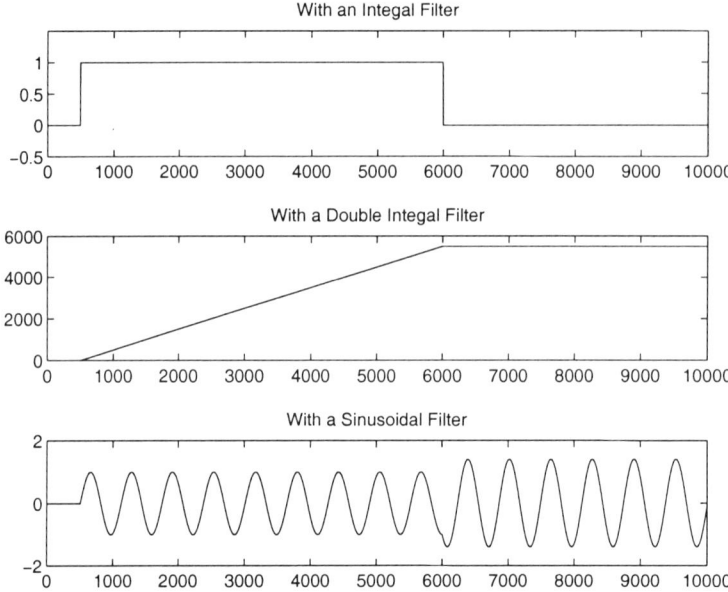

Fig. 11.3. Signal generated by passing the shock through filters

zero mean and finite variance (MacGregor, Harris, and Wright 1984) whereas, the method of formulating the disturbance model is important for controller design irrespective of the deterministic or stochastic nature of the disturbances. The minimum variance control law generally yields a minimum SSE control law[2] for deterministic disturbances (MacGregor, Harris, and Wright 1984). For example, if the disturbance model is $N = 1/(1-q^{-1})$, then the minimum variance control law will be a minimum SSE control law for step-type disturbances. Similarly, if the disturbance model is of the sinusoidal structure, then the minimum variance control law yields a minimum SSE control law for sinusoidal disturbances. These issues become more evident in the H_2 control framework in Section 11.6. Since the setpoint tracking problem can be re-formulated as a regulatory control problem, minimum variance control can handle stochastic or deterministic setpoint tracking problems naturally. Consequently, the methodology of performance assessment for stochastic regulatory control can be generalized to performance assessment of deterministic disturbance rejection as well as setpoint tracking properties of the controller, but the estimation of the performance index must be given a special treatment when disturbances are deterministic in nature. This is illustrated in the next section.

---

[2] SSE = Sum of Square Error, *i.e.*, $J = \frac{1}{M} \sum_{i=1}^{M} \epsilon_i^2$.

11.4 Performance assessment: stochastic and deterministic disturbances

We begin with an example of the difficulties in performance assessment when both stochastic and deterministic disturbances are concerned. Eriksson and Isaksson (1994) have a numerical example showing an unreasonable performance measure if minimum variance control is used as the benchmark. In our notation, the transfer functions in the example have the following values:

$$q^{-d}\tilde{T} = q^{-4}\frac{0.33}{1 - 0.67q^{-1}}$$

$$N = \frac{1 - 0.4q^{-1}}{1 - 0.67q^{-1}}$$

$$Q = \frac{0.7 - 0.47q^{-1}}{0.33 - 0.10q^{-1} - 0.23q^{-4}}$$

The white noise sequence u_t has variance $\sigma_a^2 = 0.36$. The controller is a well-designed Dahlin controller. The minimum variance can be calculated as $\sigma_{mv}^2 = 0.4033$, and the output variance with the Dahlin controller acting on the process is given as $\sigma_y^2 = 0.6115$. Hence, the performance index (in our notation) is

$$\eta_{min} \triangleq \frac{\sigma_{mv}^2}{\sigma_y^2} = \frac{0.4033}{0.6115} = 0.66$$

where η_{min} is the performance index[3] with minimum variance control as the benchmark with $0 < \eta_{min} \leq 1$. If the controller is changed to a P-only controller with a gain of 0.1745, the output variance becomes $\sigma_y^2 = 0.4037$. This yields the performance index as

$$\eta_{min} = \frac{0.4033}{0.4037} = 0.999$$

This indicates that the performance of the P-controller is better than the Dahlin controller, yet this is clearly an unacceptable conclusion. It is, however, a correct result.

Performance assessment techniques as proposed by Harris (1989) provide assessment of control loop performance under *routine operating conditions*. In this example, step-type disturbances or setpoint changes do not affect the process. An integral control is clearly not necessary in this situation. Therefore, a P-controller gives a better performance measure than the Dahlin controller which has integral action. However, integral action is practically desired in order to handle random-walk type stochastic disturbances or step-type deterministic disturbances or setpoint changes. It is necessary, therefore,

---

[3] In this chapter, the subscript "min" stands for minimum variance or optimal H_2 control. For example, η_{min} represents the performance index with minimum variance or optimal H_2 control as a benchmark.

to sample the data carefully before carrying out the performance evaluation. For example, one should ask if the set of sampled routine operating data contain effects of all the disturbances that really affect the process, *i.e.*, if the set of data is representative. Performance assessment will then estimate the benchmark which regulates the disturbances occurring during the data-sampling period optimally.

Since all disturbances can be regarded as shocks (with different probability distributions) filtered by different disturbance filters, all of the disturbances can be lumped together theoretically via the method of spectral analysis. Practically, the dynamics of the lumped disturbances can also be estimated via time series analysis. The benchmark control (one-degree-of-freedom control) would be a controller which minimizes the effect of the lumped disturbances.

11.5 Performance assessment with pure deterministic disturbances

If deterministic disturbances occur rather infrequently, e.g., when only one step-change occurs in the collected data, time series modelling of the closed-loop process cannot depict the nature of the disturbances, e.g., step-type, and performance assessment may not then be carried out under the stochastic framework. Under these circumstances, direct identification of the closed-loop transfer function from the disturbances to the process output is desired. There is no difficulty in identifying such a model if the deterministic disturbances are measurable, e.g., setpoint change, since it is equivalent to an open-loop identification problem. This model may also be identified if the deterministic output can be represented by a transfer function driven by an impulse.

Take the MIMO case with the simple interactor matrix (*i.e.*, $D = q^d I$) as an example. The closed-loop transfer function from a_t to Y_t can be written as

$$Y_t = (I + q^{-d}\tilde{T}Q)^{-1} N a_t \qquad (11.3)$$
$$= \underbrace{(F_0 + F_1 q^{-1} + \cdots + F_{d-1} q^{-(d-1)})}_{F} a_t + F_d a_{t-d} + \cdots \qquad (11.4)$$

where F_is correspond to impulse response coefficient matrices of the transfer function matrix from a_t to Y_t; and $e_t = F a_t$ is feedback invariant irrespective of the probability distribution of a_t. If a_t is a random stochastic shock, then the closed-loop transfer function matrix from a_t to Y_t can be estimated via time series analysis. If a_t is a single shock (in deterministic disturbances), then the closed-loop transfer function matrix can be identified via identification tools. In both cases the term of $e_t = F a_t$ can then be separated from the closed-loop transfer function matrix and subsequently used as a benchmark to assess control loop performance.

The performance measure for stochastic disturbances can be directly applied to performance assessment for deterministic disturbances. For stochastic disturbances, one can always normalize a_t such that $Var(a_t) = I$ by adjusting N. According to the definition of the objective function based performance index for the stochastic disturbances as defined in Chapter 8, we have

$$\eta_{min} = \frac{min(E[Y_t^T Y_t])}{E[Y_t^T Y_t]}$$

$$= \frac{tr(F_0^T F_0 + F_1^T F_1 + \cdots + F_{d-1}^T F_{d-1})}{tr(F_0^T F_0 + F_1^T F_1 + \cdots + F_{d-1}^T F_{d-1} + F_d^T F_d + F_{d+1}^T F_{d+1} + \cdots)} \quad (11.5)$$

Equation 11.5 defines an H_2 norm measure (Dahleh and Diaz-Bobillo 1995) of the system. The denominator in Equation 11.5 represents the H_2 norm of the closed-loop system, and the numerator represents the H_2 norm of the feedback controller-invariant part. From H_2 norm point of view, Equation 11.5 defines a performance measure of a deterministic system, i.e.,

$$\eta_{min} = \frac{(F_0^T F_0 + F_1^T F_1 + \cdots + F_{d-1}^T F_{d-1})}{F_0^T F_0 + F_1^T F_1 + \cdots + F_{d-1}^T F_{d-1} + F_d^T F_d + F_{d+1}^T F_{d+1} + \cdots}$$

$$= \frac{min(H_2)}{H_2}$$

11.6 Unified assessment of stochastic and deterministic systems

Whether disturbances are stochastic or deterministic in nature, they require the two-step procedure for control loop performance assessment. The first step involves filtering, i.e., time series analysis using only output data to obtain closed-loop transfer function from a_t to Y_t, in the stochastic case, and identification of the same closed-loop transfer function in the deterministic case. The second step involves calculation of the performance index, which is basically the ratio of the sum of square terms of the impulse response coefficients matrices of the closed-loop transfer function for both stochastic and deterministic disturbances. The sum of square terms of the impulse response coefficients matrices is, in fact, the H_2 norm of the system. Thus, performance assessment for both deterministic and stochastic systems may be unified under the H_2 framework.

The H_2 norm of a transfer function matrix G is defined as (Dahleh and Diaz-Bobillo 1995):

$$\|G\|_2^2 = \frac{1}{2\pi} \int_{-\pi}^{\pi} tr[G(e^{j\omega})G^T(e^{-j\omega})]d\omega$$

$$= \sum_{t=0}^{\infty} tr[F_t F_t^T]$$

where F_t is the impulse response coefficients (or Markov parameter matrices) of G. Consider the closed-loop response to disturbances a_t as shown in Figure 11.1 (with $Y_t^{sp} = 0$), which is

$$Y_t = (I + TQ)^{-1} N a_t$$

The corresponding closed-loop transfer function matrix is therefore

$$G_{cl} = (I + TQ)^{-1} N$$

If a_t is a single 'shock', then Y_t represents closed-loop response to a deterministic disturbance, and the H_2 norm of G_{cl} defines the sum of squares of the errors (SSE) of the closed-loop system. However, if a_t is a white-noise sequence with $Var(a_t) = I$, then according to Parseval's Theorem,

$$\|G_{cl}\|_2^2 = \frac{1}{2\pi} \int_{-\pi}^{\pi} tr(G_{cl}(e^{j\omega}) G_{cl}^T(e^{-j\omega})) d\omega = tr[Var(Y_t)] = E(Y_t^T Y_t)$$

which is the quadratic measure of variance of the closed-loop output. The variance matrix of the white noise sequence, a_t, can always be normalized to an identity matrix by adjusting the disturbance transfer function matrix N; therefore, the H_2 norm applies to both stochastic and deterministic systems. Consequently, the optimal H_2 control law is an optimal control law for both deterministic and stochastic disturbances. In the sequel, we simply consider the optimal or desired H_2 control as the benchmark for control loop performance assessment of both stochastic and deterministic systems. For example, the unified scalar performance index with optimal H_2 control as the benchmark is

$$\eta_{min} = \frac{min(\|G_{cl}\|_2^2)}{\|G_{cl}\|_2^2}$$

Then for stochastic disturbances, this index is precisely the objective function-based index defined in Chapter 8 for performance assessment of stochastic systems.

11.7 Simulation

For the same process, as that used in Eriksson and Isaksson (1994), performance of the Dahlin-controller and the simple P-controller is re-assessed. In addition to the "routine" disturbances from the shock a_t (normally-distributed with $Var(a_t) = 0.36$), deterministic step-type disturbances are added to the system as shown in Figure 11.4. b_t is the shock with most of its probability density concentrated on $b_t = 0$; therefore, disturbance δ_t is a randomly occurring step-type deterministic disturbance.

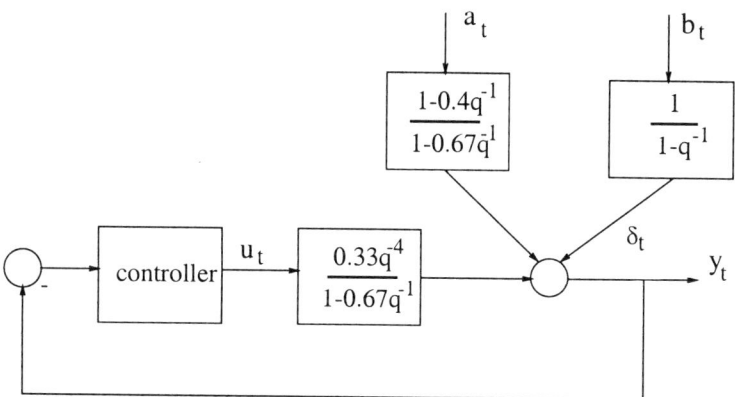

Fig. 11.4. Block diagram representation of the simulated process

Results of the process simulated with the Dahlin-controller are shown in Figures 11.5 and 11.6. The FIR model or impulse response (using 20 coefficients) of the closed-loop transfer function from δ_t to y_t is identified as

$$G_{y\delta} = 0.98 - 0.004q^{-1} + 0.057q^{-2} - 0.06q^{-3} - 0.74q^{-4} - 0.12q^{-5} - 0.09q^{-6} + \cdots$$

Owing to time delay $d = 4$, the first four terms of $G_{y\delta}$ are feedback invariant for impulse disturbances, i.e., for the disturbance model $N = 1$. For a step disturbance (i.e., for $N = 1/(1 + q^{-1})$), one can either integrate impulse response coefficients or use the output and the differenced input data directly, (i.e., filtering input data δ_t by $(1 - q^{-1})$), to obtain the step response coefficients. Direct identification (using 20 coefficients) yields the step response

$$s_t \stackrel{\triangle}{=} y_t|_{step} = 0.98 + 0.98q^{-1} + 1.03q^{-2} + 0.97q^{-3} + 0.24q^{-4} + 0.12q^{-5} + \cdots$$

The first four terms are feedback control invariant for this step-type disturbance, i.e., for $N = 1/(1 - q^{-1})$. This also implies that the peak error owing to a unit step disturbance is no less than 1 for any linear feedback controller. A comparison between theoretical step response and predicted step response with 95% error bounds (95% confidence interval) is shown in Figure 11.6. Clearly the Dahlin controller has a performance very close to minimum SSE or optimal H_2 control for the step disturbance in this example since the first four points are feedback control invariant. The performance index for the step disturbance can be calculated as

$$\begin{aligned}
\eta_{min} &= \frac{min(\|s_t\|_2^2)}{\|s_t\|_2^2} \\
&\approx \frac{0.9807^2 + 0.9775^2 + 1.0341^2 + 0.9732^2}{0.9807^2 + 0.9775^2 + 1.0341^2 + 0.9732^2 + 0.2383^2 + 0.1174^2} \\
&= 0.9824
\end{aligned}$$

11. A Unified Approach to Performance Assessment

This is very close to the index calculated theoretically which is 0.976. The residuals from Figure 11.5 can be used to assess performance of the controller for rejection of "routine" stochastic disturbances. Applying the FCOR algorithm to the residuals yields the performance index for stochastic disturbances as

$$\eta_{min} \approx 0.69$$

This result agrees with the previous analysis for the same process with only stochastic disturbances in Section 11.4. Therefore, in this example, the Dahlin controller has "optimal" performance for rejection of step-type disturbances but has relatively "low" performance for rejection of stochastic disturbances.

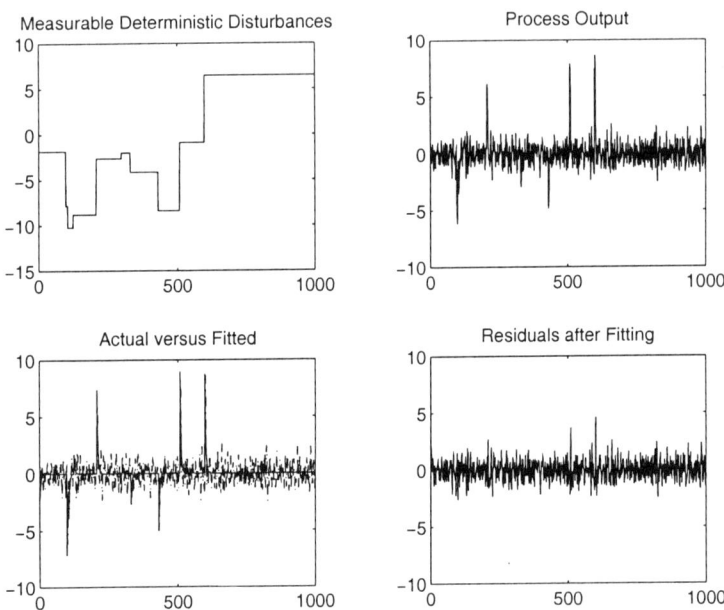

Fig. 11.5. Process response and identification results (Dahlin controller)

For the simple P-controller, using the same deterministic and stochastic disturbances as those used for the simulation of the Dahlin controller, the following polynomial transfer function is identified:

$$G_{y\delta} = \frac{0.9883 - 0.7674q^{-1}}{1 - 0.7382q^{-1}}$$

The predicted step response is

$$s_t \triangleq y_t|_{step} = \frac{0.9883 - 0.7674q^{-1}}{(1 - 0.7382q^{-1})(1 - q^{-1})}$$

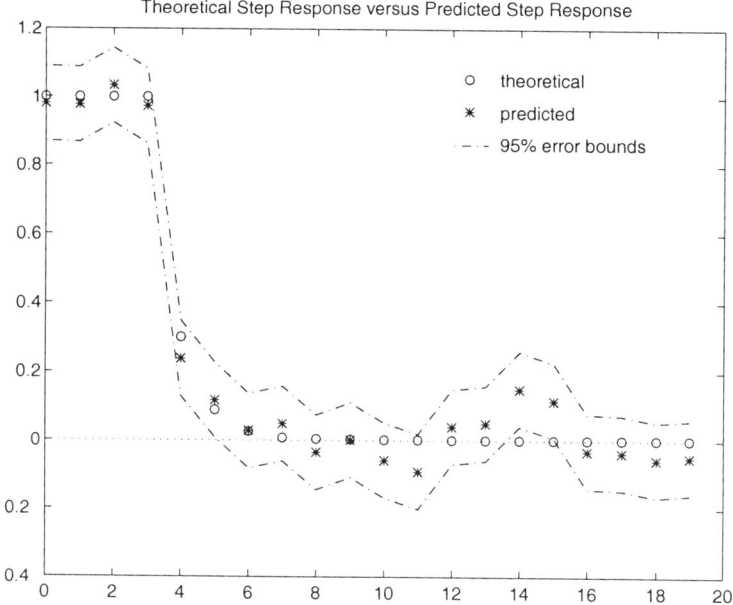

Fig. 11.6. Predicted output response to a step disturbance (Dahlin controller)

This step response has an offset. Therefore, the SSE or H_2 norm of y_t is infinite, and the performance index of the P-controller for step disturbance is

$$\eta_{min} = \frac{min(||s_t||_2^2)}{||s_t||_2^2} = 0$$

Similarly, the residuals after fitting can be used to assess performance of the P-controller for rejection of the "routine" stochastic disturbances. Applying the FCOR algorithm yields the performance index for stochastic disturbances as

$$\eta = 0.97$$

This also agrees well with the previous analysis for the process with only stochastic disturbances in Section 11.4; therefore, the simple P-controller has "optimal" performance for rejection of "routine" stochastic disturbances but has very poor performance for rejection of step-type disturbances.

11.8 Summary

Feedback control loop performance assessment for regulating both stochastic and deterministic disturbances and/or setpoint tracking has been discussed in a unified manner under the H_2 framework. It has been shown that performance assessment of deterministic disturbances and/or setpoint tracking can be treated in a very similar way to the treatment of the stochastic system. The application of the proposed method is illustrated by its evaluation on a simulated example.

CHAPTER 12
PERFORMANCE ASSESSMENT: USER-SPECIFIED BENCHMARK

12.1 Introduction

Control loop performance assessment has been extended to many situations, and many approaches have been developed as discussed in the earlier chapters, e.g., performance assessment of: 1) SISO feedback control systems (Desborough and Harris 1992; Stanfelj, Marlin, and MacGregor 1993; Kozub and Garcia 1993; Lynch and Dumont 1993; Tyler and Morari 1996); 2) feedback control of nonminimum phase SISO systems (Tyler and Morari 1995); 3) MIMO feedback control systems (Huang, Shah, and Kwok 1995; Huang, Shah, and Kwok 1996; Harris, Boudreau, and MacGregor 1995; Harris, Boudreau, and MacGregor 1996). The portion of a process output that is feedback controller invariant determines the minimum variance achievable theoretically and characterizes the most fundamental performance limitation of a system owing to the existence of time-delays or infinite zeros. Practically, however, there are many other limitations on the achievable control loop performance. The existence of nonminimum phase or poorly damped zeros, sampling rate, amplitude and/or rate constraints on control action, robustness constraints *etc.* are examples of such limitations; therefore, a feedback controller that indicates performance reasonably close to minimum variance control does not require further tuning (if the variance is the most important issue) while a feedback controller that indicates poor performance relative to minimum variance control is not necessarily a poor controller. Further analysis of performance limitations and comparison with more realistic benchmarks are usually required. Performance assessment with minimum variance control as a benchmark requires minimum effort (routine operating data plus *a priori* knowledge of time-delays), and serves as the most convenient first-level performance assessment test, therefore, (if the variance is the main point of interest). Only those loops that indicate poor first-level performance need to be re-evaluated by higher-level performance assessment tests. A higher-level performance test usually requires more *a priori* knowledge than just a knowledge of time-delays. This chapter addresses practical issues that are of interest for such a higher-level performance assessment test.

The main contribution of this chapter is to propose a technique of practical control loop performance assessment relative to a benchmark in terms of user-specified closed-loop dynamics. All of these are discussed for SISO and

MIMO systems. Normally, the MIMO case is general and includes the SISO system as a special case, but the delay matrix (or the interactor matrix) of the MIMO system is not a simple extension to the time-delay term of the SISO system. Thus, for clarity of presentation, the SISO case is considered first. This chapter is organized as follows: Performance assessment of minimum phase systems with desired closed-loop dynamics as a more practical benchmark is considered in Section 12.3, and the treatment of nonminimum phase systems is discussed in Section 12.4, followed by concluding remarks in Section 12.5.

12.2 Preliminaries

In Section 12.3, we will assume that the plant transfer function matrix T has no zeros or poles outside the unit circle except for the time-delays (infinite zeros). This condition will be relaxed in Section 12.4. The disturbance transfer function matrix N has no poles outside the unit circle but can have zeros outside the unit circle. However, an all-pass factor N_p, which contains all zeros that are outside the unit circle including the infinite zeros, can be factored out from N, such that $N = \tilde{N}N_p$ and $\tilde{N}$ is a minimum-phase transfer function matrix. This factorization does not change the H_2 norm of the closed-loop system, i.e., $||(I + TQ)^{-1}N||_2^2 = ||(I + TQ)^{-1}\tilde{N}||_2^2$. The admissible minimum variance or optimal H_2 control law does not depend on the all-pass term N_p(Astrom and Wittenmark 1990; Morari and Zafiriou 1989); therefore, if there are unstable or infinite zeros in N, one simply needs to factorize an all-pass term from it. The remaining $\tilde{N}$ term is then considered as the disturbance transfer function matrix. This will not affect the H_2 norm of the closed-loop system and its optimal control law. Time series analysis automatically produces such minimum phase disturbance transfer function matrix. Thus, we assume that the disturbance transfer function matrix N has no zeros outside the unit circle as well.

12.3 User-specified performance benchmark: minimum phase systems

12.3.1 SISO case

The minimum variance or optimal H_2 control law serves as a good global reference point to assess control loop performance, but the minimum variance or optimal H_2 control law may not be the one desired in practice. For example, if the process has a fast controller sampling rate, then minimum variance or minimum SSE control with such a sampling rate usually requires excessive control actions; therefore, in many practical circumstances, a more realistic

12.3 User-specified performance benchmark: minimum phase systems

user-specified benchmark control is desirable. For example, one may wish to consider desired closed-loop dynamics as a reference benchmark in terms of settling times, overshoot *etc.* Specifically it would be of interest to know if the actual closed-loop dynamics are close to or far away from the desired dynamics. Kozub and Garcia(1993) have suggested that one of the alternatives for the practical benchmark performance can be the minimum variance control response (finite moving-average term) filtered by a filter. Tyler and Morari(1995b) suggested using a generalized likelihood ratio to test whether the actual performance (in the form of impulse response coefficients of the closed-loop transfer function $G_{cl} = (I + TQ)^{-1}N$) is in the set of the desired performance. All of these are limited to SISO applications, but since the actual performance (in the form of impulse response coefficients of G_{cl}) can be estimated from data using time series analysis or standard identification tools (Ljung 1998), performance assessment can be directly computed by comparing the actual impulse response coefficients with those desired. Direct observation and analysis of the actual impulse response coefficients gives one significantly more information than a simple "yes or no" test, and can be easily extended to MIMO applications. For example, a maximum likelihood ratio test cannot tell whether the controller is over-tuned, under-tuned, or in what way it is different from the desired performance. On the other hand, a cursory study of the impulse response coefficients may easily provide such tuning guidelines.

An important fact that has been ignored by other researchers is that the desired closed-loop dynamics (G_{des}) cannot be specified arbitrarily. They need to consider the physical limitations. For example, closed-loop response within the time-delay period is feedback control invariant and cannot be specified by users. Nonminimum phase zeros cannot be canceled by a stable controller and must affect the desired closed-loop dynamics as well. These limitations are considered in the present chapter when the desired closed-loop dynamics are specified.

The differences between optimal H_2 control and the user specified benchmark control (desired closed-loop dynamics) can be clearly seen from Equation 11.4. For optimal H_2 control, the remaining terms after the first d terms should be zero. For user specified benchmark control, the first d terms should be the same as those under optimal H_2 control, but the remaining terms are no longer zero. These remaining terms define the desired closed-loop dynamics; therefore, the closed-loop dynamics, e.g., for the SISO case, have the following form[1]:

$$y_t|_{user} = \underbrace{(f_0 + f_1 q^{-1} + \cdots + f_{d-2} q^{-d+2} + f_{d-1} q^{-d+1}}_{F} + q^{-d} G_R) a_t$$

(12.1)

---

[1] In this chapter, the subscript "user" stands for a user-specified benchmark control. For example, y_{user} represents the process output under the user-specified control.

where G_R is a stable and proper transfer function. There are many ways to specify the term G_R based on information, such as closed-loop settling time, time constant, decay ratio, desired variance, frequency domain characteristics, robust performance etc. If G_R is directly specified as desired closed-loop dynamics, i.e., $G_R = G_{des}$, then only *a priori* knowledge of time-delays is required for the calculation of such benchmark performance. If the desired closed-loop response is specified by some other characteristic, such as settling time, then G_R consists of a set of transfer functions and no explicit expression for G_R is actually defined. One can simply test, for example, whether the actual closed-loop settling time is the same as the desired value. However, all these specifications of G_R are arbitrary in some way, and it is not clear how such specifications affect closed-loop dynamics, e.g., in terms of performance optimization and robustness. For example, in the specification of the settling time, there is an infinite number of G_Rs that can be considered, and one does not know which one has the performance closest to optimal control. On the other hand, it is well-known that optimal H_2 control augmented by a filter improves robust performance and provides a good compromise between performance and robustness; in addition the closed-loop dynamics can be adjusted by the tuning of the filter parameters (Mohtadi 1988; Morari and Zafiriou 1989).

Consider the specification of G_R as

$$G_R = (1 - G_F)R \tag{12.2}$$

where G_F is a stable and proper filter and, R (a rational proper transfer function) is defined via the Diophantine identity:

$$N = F + q^{-d}R$$

Then Equation 12.1 becomes

$$y_t|_{user} = (f_0 + f_1 q^{-1} + \cdots + f_{d-1} q^{-d+1} + q^{-d}(1 - G_F)R) a_t \tag{12.3}$$

The filter G_F can be specified according to the desired closed-loop dynamics. It should be chosen in such a way that $(1 - G_F)R$ converges to zero asymptotically, i.e., no offset occurs. For example, if R has a pole equal to 1, e.g., step-type disturbances, then $(1 - G_F)$ must have a zero equal to 1 in order to preserve the asymptotic property of the optimal H_2 control law. If a commonly used first-order filter, which satisfies the asymptotic property for type 1 system, is specified as

$$G_F = \frac{1-\alpha}{1-\alpha q^{-1}}$$

then α can be calculated via the desired closed-loop settling time or time constant by

$$\alpha = exp(-\frac{\Delta T}{\tau})$$

12.3 User-specified performance benchmark: minimum phase systems

where ΔT is the sampling interval and τ is the desired time constant of the closed-loop process. However, one should note that the closed-loop dynamics also depend on R in addition to the filter dynamics.

We shall show that the specification of the closed-loop system, as in Equation 12.1, is practically achievable for a minimum-phase system, and the specification in Equation 12.2 is equivalent to the H_2 optimal control law augmented by a filter.

Consider the controller specification in the IMC framework as shown in Figure 12.1. Write the plant transfer function as $T = q^{-d}\tilde{T}$, where $\tilde{T}$ is the delay-free transfer function. Then by assuming $\hat{T} = T$, we have

$$\begin{aligned} y_t &= (1 - q^{-d}\tilde{T}Q^*)Na_t \\ &= Na_t - q^{-d}\tilde{T}Q^*Na_t \\ &= (F + q^{-d}R)a_t - q^{-d}\tilde{T}Q^*Na_t \\ &= Fa_t + q^{-d}(R - \tilde{T}Q^*N)a_t \end{aligned} \quad (12.4)$$

Equating 12.4 and 12.1 yields

$$R - \tilde{T}Q^*N = G_R$$

This results in

$$Q^* = \frac{R - G_R}{\tilde{T}N} \quad (12.5)$$

This IMC controller is proper and stable. Therefore, the closed-loop response specified in Equation 12.1 is achievable. For specification in Equation 12.2, Equation 12.5 becomes

$$Q^* = \frac{R - (1 - G_F)R}{\tilde{T}N} = \frac{G_F R}{\tilde{T}N} \quad (12.6)$$

If $G_F = 0$, then $Q^* = 0$ and the controller is in open-loop mode. If $G_F = 1$, then according to Equation 12.3,

$$y_t|_{user} = (f_0 + f_1 q^{-1} + \cdots + f_{d-1} q^{-(d-1)})a_t$$

This is the minimum variance control response or optimal H_2 control response. Thus, the controller, as specified in Equation 12.6, is in fact an optimal H_2 control law augmented by a filter G_F (Morari and Zafiriou 1989). The role of the filter is to adjust or tune the controller from the open-loop mode ($G_F = 0$) to optimal H_2 control mode ($G_F = 1$).

The specification in Equation 12.2 requires a knowledge of R. The term R must be calculated from the disturbance transfer function N via the Diophantine identity so the disturbance transfer function should be known in order to apply such a specification. For most setpoint tracking problems and some regulatory problems, the dynamics of the setpoint are known as *a priori* knowledge. For example, if one is interested in the tracking performance of a step-type setpoint change or regulatory performance of step-type disturbances, then the disturbance dynamics are simply $N = 1/(1 - q^{-1})$. In Section 12.4, it will be shown that, even if *a priori* knowledge of N is not available, the disturbance transfer function N can be estimated conveniently with dither excitation under closed-loop conditions.

Since complete dynamics of the user-specified closed-loop system are available, one can compare the current closed-loop dynamics with the user-specified closed-loop dynamics directly. For example, to calculate the performance index under the stochastic framework, the variance of the user-specified benchmark control can be calculated from Equation 12.1 and is defined as σ^2_{user}. The performance index can then be calculated as the ratio:

$$\eta_{user} = \frac{\sigma^2_{user}}{\sigma^2_y}$$

Example 12.3.1. Consider a first-order process with a time-delay and the transfer function represented by

$$T = \frac{(2 - q^{-1})q^{-2}}{1 - 0.8q^{-1}}$$

Let the disturbance transfer function be

$$N = \frac{1}{1 - q^{-1}}$$

The sampling interval $\Delta T = 5$ sec. The main objective in this design is to regulate y_t in the presence of integrated white noise or random-walk type disturbances. The choice of the desired closed-loop time constant, $0.5 \leq \tau \leq 1$ (minute), reflects this. If setpoint tracking performance was of interest, perhaps a smaller desired closed-loop time constant could have been specified.

Since the time-delay $d = 2$, N is expanded according to the Diophantine identity as

$$N = \underbrace{1 + q^{-1}}_{F} + q^{-2} \underbrace{\frac{1}{1 - q^{-1}}}_{R}$$

Therefore, the minimum variance is

$$\sigma^2_{mv} = Var(Fa_t) = 2\sigma^2_a$$

Now consider a first-order filter:

$$f = \frac{1 - \alpha}{1 - \alpha q^{-1}}$$

To satisfy the desired closed-loop time constant $0.5 \leq \tau \leq 1$ (minute) with sampling interval $\Delta T = 5$ s, the parameter α can be calculated from

12.3 User-specified performance benchmark: minimum phase systems

$$\alpha = exp(-\frac{\Delta T}{\tau})$$

This results in $0.85 \leq \alpha \leq 0.92$. The desired closed-loop response can be calculated from Equation 12.3 as

$$\begin{aligned} y_t|_{user} &= f_0 a_t + f_1 a_{t-1} + (1 - \frac{1-\alpha}{1-\alpha q^{-1}})\frac{1}{1-q^{-1}} a_{t-2} \\ &= a_t + a_{t-1} + \frac{\alpha}{1-\alpha q^{-1}} a_{t-2} \quad (12.7) \\ &= \frac{1 + (1-\alpha)q^{-1}}{1 - \alpha q^{-1}} a_t \quad (12.8) \end{aligned}$$

where $0.85 \leq \alpha \leq 0.92$. This gives the achievable user-specified closed-loop response. The variance of the desired closed-loop can also be calculated from Equation 12.8 as

$$\sigma_{user}^2|_{\alpha=0.85} = Var(y_t)|_{\alpha=0.85} = 4.6036\sigma_a^2$$

and

$$\sigma_{user}^2|_{\alpha=0.92} = Var(y_t)|_{\alpha=0.92} = 7.5104\sigma_a^2$$

Therefore,

$$4.6036\sigma_a^2 \leq \sigma_{user}^2 \leq 7.5104\sigma_a^2$$

Now consider the same process under integral control

$$Q = \frac{0.1}{1 - q^{-1}}$$

with $Var(a_t) = 1$. The closed-loop system was simulated and 5000 data points were recorded. A truncated moving-average model is obtained from time series analysis of y_t. The coefficients of this moving-average model correspond to the closed-loop impulse response coefficients. These coefficients together with their 95% bounds are plotted in Figure 12.2. The desired closed-loop dynamics have been shown in Equation 12.8. Their corresponding impulse response coefficients are calculated and plotted together with the actual impulse response coefficients in Figure 12.2. Since the desired impulse response is not a single curve but a region (a region between the two solid lines), any actual impulse response which falls within this region is considered acceptable; however, the actual impulse response coefficients are estimated values so a statistical test should be used to determine whether or not the actual impulse response falls into this region. One can consider the desired performance region as if it constitutes a thick or 'fuzzy' desired impulse response curve: 1) for any particular actual impulse response coefficient, if its 95% confidence interval and the 'fuzzy' desired performance region do not intersect, then we may conclude that this particular coefficient does not fall in the desired region with 95% confidence; 2) if more than 5% of impulse response coefficients, over the time-period of interest, do not fall in the desired region as tested in the first step, then one may conclude that the actual performance does not lie in the set of the desired performance. The confidence bounds can always be narrowed by increasing the data sampling size, and, hence, the reliability of such tests can be increased. This is a quantitative or 'yes' or 'no' decision criterion, but a visual or qualitative analysis of the plot is more important and it is recommended that this be performed. Figure 12.2 clearly shows unacceptable performance of the integral controller. The actual closed-loop behaves as an underdamped system, and, in fact, the system appears to be over-tuned.

Now we add a second-order filter to the integral controller:

$$Q = (\frac{0.05 - 0.04q^{-1}}{2 - 0.9q^{-1} - 0.05q^{-2}})\frac{1}{1 - q^{-1}}$$

The simulation result is shown in Figure 12.3. The actual closed-loop dynamics are slightly slower than the desired response, and, in fact, are not in the desired set. The controller appears to be under-tuned. Finally, a slightly aggressive controller

$$Q = (\frac{0.11 - 0.088q^{-1}}{2 - 0.78q^{-1} - 0.11q^{-2}})\frac{1}{1 - q^{-1}}$$

is implemented. The result shown in Figure 12.4 indicates that the actual performance is now in the set of the desired performance.

Note that by using the same method, one can make many performance tests. For example, one can even specify the desired performance set in terms of the frequency domain specifications. The actual closed-loop dynamics together with their confidence intervals in the frequency domain can then be calculated. The performance test can be made in the frequency domain. The significance of this method is that it can provide information including tuning guidelines via observation of the patterns of the impulse response coefficients or the spectrum of the closed-loop systems.

12.3.2 MIMO case

Consider the MIMO system:

$$Y_t = TU_t + Na_t$$

where T is a process transfer function matrix, N is the disturbance transfer function matrix, Y_t, U_t and a_t are output, input and disturbance with appropriate dimensions. A unitary interactor matrix D, with $D^T(q^{-1})D(q) = I$, can be factored from the transfer function matrix T, such that $\tilde{T} = DT$, which is a delay free transfer function matrix. The factorization of such a unitary interactor matrix does not change the H_2 norm of the transfer function matrix, i.e., $||\tilde{T}||_2^2 = ||T||_2^2$. Using the Diophantine identity, the disturbance transfer function matrix N can be expanded as

12.3 User-specified performance benchmark: minimum phase systems   141

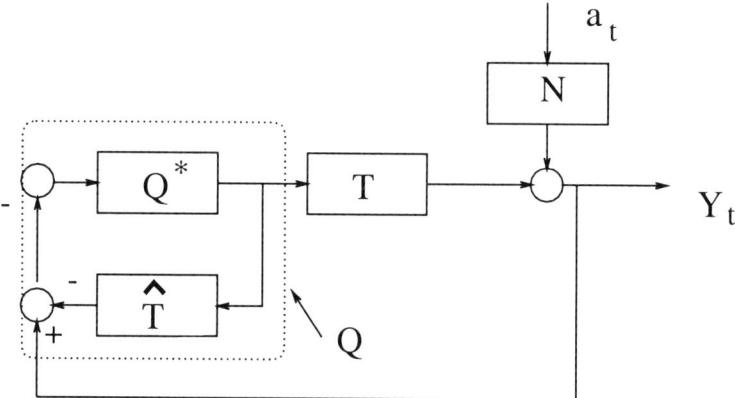

Fig. 12.1. Control loop configuration under the IMC framework

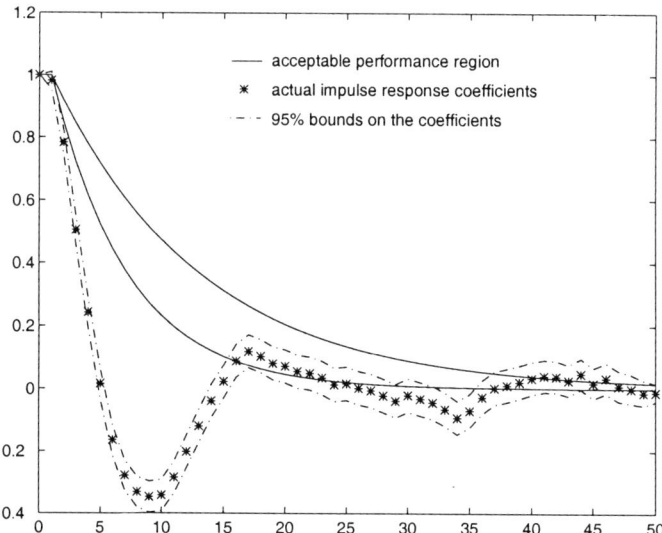

Fig. 12.2. Closed-loop impulse response coefficients for a simple integral controller.

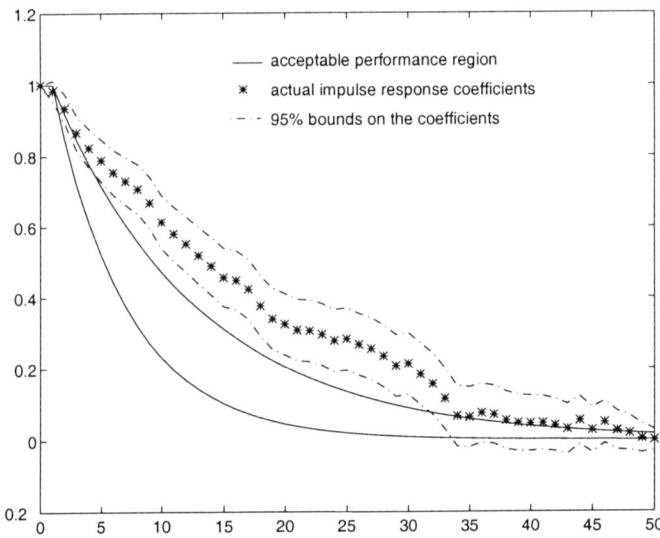

Fig. 12.3. Closed-loop impulse response coefficients for an integral plus filter controller.

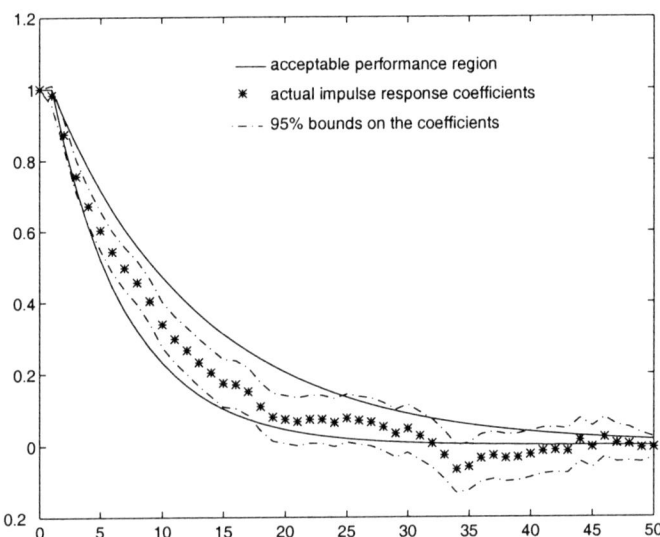

Fig. 12.4. Closed-loop impulse response coefficients for a detuned integral plus filter controller.

12.3 User-specified performance benchmark: minimum phase systems

$$q^{-d}DN = \underbrace{F_0 + F_1 q^{-1} + \cdots + F_{d-1} q^{-(d-1)}}_{F} + q^{-d} R$$

where F_i (for $i = 1, \cdots, d-1$) are constant coefficient matrices, and R is the remaining proper transfer function matrix after the expansion. In Chapter 8 the minimum variance or optimal H_2 norm control response has been shown to be[2]:

$$Y_t|_{min} = q^d D^{-1}(F_0 + F_1 q^{-1} + \cdots + F_{d-1} q^{-(d-1)}) a_t \qquad (12.9)$$
$$= (E_0 + E_1 q^{-1} + \cdots + E_{d-1} q^{-(d-1)}) a_t \qquad (12.10)$$

where E_i (for $i = 1, \cdots, d-1$) are constant coefficient matrices. It has also been shown in Chapter 8 that the minimum variance control response $Y_t|_{min}$ can be estimated from routine operating data with *a priori* knowledge of the unitary interactor matrix D.

For performance assessment with a reference benchmark different from minimum variance or optimal H_2 norm control, Equation 12.10 should be augmented with a user-specified transfer function matrix G_R

$$Y_t|_{user} = (E_0 + E_1 q^{-1} + \cdots + E_{d-1} q^{-(d-1)} + G_R q^{-d}) a_t \qquad (12.11)$$

G_R is a stable and proper transfer function matrix. In practice, G_R may be specified as a diagonal matrix with the dynamics of each output corresponding to the diagonal elements. Now we are in the position to show that the user-specified closed-loop response, as shown in Equation 12.11, is achievable.

Under the IMC framework, the closed-loop response can be written as

$$Y_t = (I - TQ^*) N a_t$$

Factor T as $T = D^{-1}\tilde{T}$, where $\tilde{T}$ is the delay-free transfer function matrix. Then

$$Y_t = (I - D^{-1}\tilde{T}Q^*) N a_t$$
$$= q^d D^{-1}(q^{-d}DN - q^{-d}\tilde{T}Q^* N) a_t$$

Using the Diophantine identity $q^{-d}DN = F + q^{-d}R$, we have

$$Y_t = q^d D^{-1}(F + q^{-d}R - q^{-d}\tilde{T}Q^* N) a_t$$
$$= q^d D^{-1} F a_t + D^{-1}(R - \tilde{T}Q^* N) a_t \qquad (12.12)$$

As shown in Chapter 8, the first term on the right hand side of Equation 12.12 is the minimum variance control response and can be written as

$$Y_t|_{min} = q^d D^{-1} F a_t = (E_0 + \cdots + E_{d-1} q^{-d+1}) a_t \qquad (12.13)$$

---

[2] In Chapter 8, $Y_t|_{mv}$ represents the process output under minimum variance control. In this chapter, this term is changed to $Y_t|_{min}$ to reflect the minimum variance or equivalently optimal H_2 control.

Substituting Equation 12.13 into Equation 12.12, and then equating the result to Equation 12.11 yields

$$D^{-1}(R - \tilde{T}Q^*N) = G_R q^{-d}$$

This results in

$$q^d D^{-1}(R - \tilde{T}Q^*N) = G_R \tag{12.14}$$

Solving Equation 12.14 yields

$$Q^* = \tilde{T}^{-1}(R - q^{-d}DG_R)N^{-1}$$

Again, Q^* is proper and stable, and is, therefore, an achievable controller.

As discussed in the SISO case, there are many ways to specify the transfer function matrix G_R. One may directly specify it as a desired transfer function matrix, i.e., $G_R = G_{des}$. Performance assessment with such specification requires only routine operating data plus *a priori* knowledge of the interactor matrix.

Alternatively, one may add a desired term into Equation 12.9 directly such that

$$Y_t|_{min} = q^d D^{-1}(F_0 + F_1 q^{-1} + \cdots + F_{d-1} q^{-(d-1)} + q^{-d} G_R) a_t$$

where $G_R = I - G_F$, and G_F is the user specified filter transfer function matrix according to the desired closed-loop dynamics. The filter serves as a tuning knob between optimal H_2 control ($G_F = I$) and open-loop performance ($G_F = 0$).

Example 12.3.2. Consider a process with the following transfer function matrices:

$$T = \begin{bmatrix} \frac{q^{-1}}{1-0.4q^{-1}} & \frac{q^{-2}}{1-0.1q^{-1}} \\ \frac{0.3q^{-1}}{1-0.1q^{-1}} & \frac{q^{-2}}{1-0.8q^{-1}} \end{bmatrix}$$

$$N = \begin{bmatrix} \frac{1}{1-0.5q^{-1}} & \frac{-0.6}{1-0.5q^{-1}} \\ \frac{0.5}{1-0.5q^{-1}} & \frac{1.0}{1-0.5q^{-1}} \end{bmatrix}$$

A unitary interactor matrix D can be factored out as

$$D = \begin{bmatrix} -0.9578q & -0.2873q \\ -0.2873q^2 & 0.9578q^2 \end{bmatrix}$$

Then

$$q^{-d}DN = \underbrace{\begin{bmatrix} -1.1014q^{-1} & 0.2874q^{-1} \\ 0.1916 + 0.0958q^{-1} & 1.1302 + 0.5651q^{-1} \end{bmatrix}}_{F} +$$

$$+ q^{-2} \underbrace{\begin{bmatrix} \frac{-0.5507}{1-0.5q^{-1}} & \frac{0.1437}{1-0.5q^{-1}} \\ \frac{0.0479}{1-0.5q^{-1}} & \frac{0.2826}{1-0.5q^{-1}} \end{bmatrix}}_{R}$$

12.3 User-specified performance benchmark: minimum phase systems

The minimum variance term can be written as

$$e_t = Fa_t = \begin{bmatrix} -1.1014q^{-1} & 0.2874q^{-1} \\ 0.1916 + 0.0958q^{-1} & 1.1302 + 0.5651q^{-1} \end{bmatrix} a_t$$

Therefore,

$$Y_t|_{min} = q^d D^{-1} e_t$$

$$= (\underbrace{\begin{bmatrix} 1 & -0.6 \\ 0.5 & 1 \end{bmatrix}}_{E_0} + q^{-1} \underbrace{\begin{bmatrix} -0.0275 & -0.1624 \\ 0.0918 & 0.5413 \end{bmatrix}}_{E_1}) a_t$$

Note that this explicit expression for $Y_t|_{min}$ can always be estimated from routine operating data under any feedback control with *a priori* knowledge of the unitary interactor matrix. If we assume $\Sigma_a = Ea_t a_t^T = I$, then the minimum variance can be calculated as

$$\Sigma_{min} = Var(Y_t|_{min}) = E_0 E_0^T + E_1 E_1^T = \begin{bmatrix} 1.3871 & -0.1904 \\ -0.1904 & 1.5504 \end{bmatrix}$$

with the quadratic performance measure (H_2 norm) as

$$E(Y_t^T Y_t)|_{min} = tr(\Sigma_{min}) = 2.9375$$

Now we assume that the controller sampling interval is $\Delta T = 5$ s, the desired response of output #1 is first-order with time constant $\tau_1 = 1$ min, and the desired response of output #2 is also first-order but with time constant $\tau_2 = 0.5$ min. If the desired closed-loop response is specified as

$$Y_t|_{user} = (E_0 + E_1 q^{-1} + \cdots + G_F E_{d-1} q^{-(d-1)}) a_t \quad (12.15)$$

i.e., the closed-loop impulse response coefficient matrices decay steadily at the desired time constants starting from the last feedback control invariant term, E_{d-1}, then the filter transfer function matrix should be designed according to the desired dynamics as

$$G_F = \begin{bmatrix} \frac{1}{1-0.92q^{-1}} & 0 \\ 0 & \frac{1}{1-0.85q^{-1}} \end{bmatrix}$$

Equation 12.15 can be further written as

$$Y_t = (E_0 + E_1 q^{-1} + \cdots + E_{d-1} q^{-(d-1)} + \underbrace{q(G_F - I) E_{d-1} q^{-d}}_{G_R}) a_t$$

Since $G_F(q^{-1} = 0) = I$, $G_R = q(G_F - I) E_{d-1}$ is proper. According to Equation 12.11, this is an achievable closed-loop response. Substituting numerical values to Equation 12.15 yields

$$Y_t|_{user} = \begin{bmatrix} \frac{1-0.9475q^{-1}}{1-0.92q^{-1}} & \frac{-0.6+0.3896q^{-1}}{1-0.92q^{-1}} \\ \frac{0.5-0.3332q^{-1}}{1-0.85q^{-1}} & \frac{1-0.3087q^{-1}}{1-0.85q^{-1}} \end{bmatrix} a_t$$

This achievable closed-loop response satisfies the user requirement, and can be estimated from routine operating data under any feedback control with a *priori* knowledge of the unitary interactor matrix. Its variance can be calculated as

$$\Sigma_{user} = Var(Y_t|_{user}) = \begin{bmatrix} 1.5366 & -0.5148 \\ -0.5148 & 2.3362 \end{bmatrix}$$

with the quadratic performance measure (H_2 norm) as

$$E(Y_t^T Y_t)|_{user} = tr(\Sigma_{user}) = 3.8729$$

which is 1.3 times as large as $E(Y_t^T Y_t)|_{min}$.

12.4 User-specified performance benchmark: nonminimum phase systems

In this section, we relax the assumption of minimum-phase plants by assuming that the plant transfer function matrix T can have zeros outside the unit circle. The user-specified performance assessment of the MIMO process when the process has non-minimum phase zeros is discussed in this section.

Consider the MIMO system:

$$Y_t = TU_t + Na_t$$

Factor T as $T = D_G^{-1}\tilde{T}$, where D_G^{-1} is the all-pass factor or the generalized unitary interactor matrix which contains both infinite and non-minimum phase zeros of T as discussed in Chapter 10. The generalized unitary interactor matrix D_G^{-1} can also be regarded as an all-pass factor of T, and can be calculated via the Inner-Outer factorization (Chu 1985) as well. A proper feedback controller cannot cancel time-delays. Time-delays are therefore constraints on the achievable performance. A stable controller cannot cancel non-minimum phase zeros. Non-minimum phase zeros also impose constraints on the achievable performance.

In Chapter 10, we have shown a procedure for performance assessment of nonminimum-phase MIMO processes. The procedure consists of the following steps:

1. Factor the generalized unitary interactor matrix from T as D_G;
2. Fit routine operating data Y_t by a time series model, G_{cl};
3. Multiply G_{cl} by $q^{-d}D_G$ to obtain $G'_{cl} = q^{-d}D_G G_{cl}$, where d is the order of the interactor matrix;

12.4 User-specified performance benchmark: nonminimum phase systems

4. Expand G'_{cl} into

$$G'_{cl} = \underbrace{F_0 + F_1 q^{-1} + \cdots + F_{d-1} q^{-(d-1)}}_{F} + q^{-d}\phi$$

where F_is ($i = 1, 2, \cdots, d-1$) are constant coefficient matrices, and ϕ is the remaining (proper) term after the expansion;

5. Using partial fraction expansion, ϕ can be expanded into

$$\phi = R_{nmp} + L$$

where all poles of R_{nmp} are unstable poles which are the non-minimum phase zeros of T, and L is the remaining term after the partial fraction expansion. Then the process under admissible minimum variance or optimal H_2 control can be written as[3]

$$Y_t|_{admv} = q^d D_G^{-1}(F + q^{-d} R_{nmp}) a_t \qquad (12.16)$$

This procedure will evaluate control performance with the admissible minimum variance or optimal H_2 control as the benchmark. To assess control performance with the user specified benchmark as the desired closed-loop dynamics, Equation 12.16 should be augmented with a user specified transfer function matrix G_R

$$Y_t|_{user} = q^d D_G^{-1}(F + q^{-d} R_{nmp} + q^{-d} G_R) a_t \qquad (12.17)$$

Note that the closed-loop dynamics also depend on the poles of D_G^{-1} (reciprocals of the non-minimum phase zeros) in addition to the poles of G_R.

Now it can be shown that the specification in Equation 12.17 gives a performance that is achievable. Under the IMC framework, closed-loop response can be written as

$$\begin{aligned} Y_t &= (I - TQ^*) N a_t \\ &= N a_t - TQ^* N a_t \qquad (12.18) \\ &= (q^d D_G^{-1})(q^{-d} D_G N) a_t - TQ^* N a_t \qquad (12.19) \end{aligned}$$

Using the Diophantine identity, $q^{-d} D_G N$ can be expanded as

$$q^{-d} D_G N = \underbrace{F_0 + F_1 q^{-1} + \cdots + F_{d-1} q^{-(d-1)}}_{F} + q^{-d} \underbrace{(R_{nmp} + R_{mp})}_{R}$$

where R_{nmp} contains all unstable poles of R after partial fraction expansion, while R_{mp} contains all stable poles after the partial fraction expansion. Using this identity, Equation 12.19 can be written as

---

[3] In this chapter, the subscript "admv" stands for "admissible minimum variance control". For example, $Y_t|_{admv}$ represents the process output under the admissible minimum variance control.

$$Y_t = q^d D_G^{-1}(F + q^{-d}R_{nmp} + q^{-d}R_{mp})a_t - TQ^*Na_t$$
$$= q^d D_G^{-1}(F + q^{-d}R_{nmp})a_t + [(q^d D_G^{-1})R_{mp}q^{-d} - TQ^*N]a_t \quad (12.20)$$

Equating Equation 12.20 and Equation 12.17 yields

$$(q^d D_G^{-1})R_{mp}q^{-d} - TQ^*N = q^d D_G^{-1}G_R q^{-d}$$

This can be written as

$$(q^d D_G^{-1})R_{mp}q^{-d} - D_G^{-1}\tilde{T}Q^*N = q^d D_G^{-1}G_R q^{-d} \quad (12.21)$$

Solving Equation 12.21 yields

$$Q^* = \tilde{T}^{-1}(R_{mp} - G_R)N^{-1} \quad (12.22)$$

This is an achievable IMC control law.

As in the discussion for the minimum phase system, if one specifies $G_R = (I - G_F)R_{mp}$, where G_F is a filter transfer function matrix, then Equation 12.22 becomes

$$Q^* = \tilde{T}^{-1}G_F R_{mp}N^{-1} \quad (12.23)$$

and Equation 12.17 becomes

$$Y_t|_{user} = q^d D_G^{-1}(F + q^{-d}R_{nmp} + q^{-d}(I - G_F)R_{mp})a_t \quad (12.24)$$

If $G_f = 0$, then the process is in the open-loop mode. If $G_F = I$, then Equation 12.24 is the admissible minimum variance or optimal H_2 control response. The filter G_F adjusts the controller performance between open-loop mode and optimal H_2 control.

Consider the closed-loop transfer function defined by

$$G_w = (I + TQ)^{-1}T$$

G_w is the closed-loop transfer function from the dither signal w_t to the output Y_t (Figure 12.5). With dither signal excitation, G_w can be identified under closed-loop conditions. Identification of the closed-loop transfer function matrix G_w under closed-loop conditions is equivalent to an open-loop identification problem.

Using routine operating data, the disturbance transfer function N can be identified via time series analysis of sensitivity-filtered data $S^{-1}Y_t$, i.e.,

$$S^{-1}Y_t = (I + TQ)Y_t = Na_t$$

where the sensitivity function S is defined by $S = (I + TQ)^{-1}$. Since $G_w = (I + TQ)^{-1}T$, we have

$$S = I - G_w Q$$

12.4 User-specified performance benchmark: nonminimum phase systems

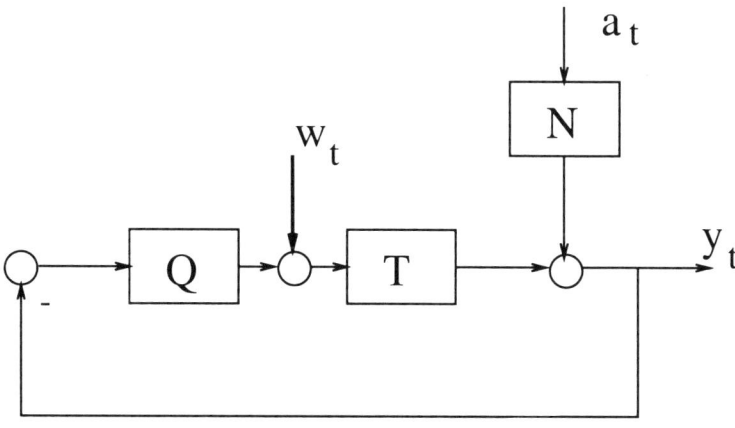

Fig. 12.5. Block diagram of the closed-loop system.

where Q is normally assumed to be known, but can also be directly identified from closed-loop data; therefore, the disturbance transfer function matrix can be estimated under closed-loop conditions. With the knowledge of D_G, N and the user specified filter G_F, one can calculate the user specified benchmark performance via Equation 12.24. The simplest form of G_R will be a diagonal matrix.

Example 12.4.1. Consider the following system from Tsiligiannis and Svoronos (1989):

$$T = \begin{bmatrix} \frac{0.6q^{-1}}{1-0.4q^{-1}} & \frac{0.5q^{-1}}{1-0.5q^{-1}} \\ \frac{0.6q^{-1}}{1-0.5q^{-1}} & \frac{0.6q^{-1}}{1-0.4q^{-1}} \end{bmatrix}$$

Assume the disturbance transfer function matrix as

$$N = \begin{bmatrix} \frac{1}{1-0.5q^{-1}} & \frac{-0.6}{1-0.5q^{-1}} \\ \frac{0.5}{1-0.5q^{-1}} & \frac{1}{1-0.5q^{-1}} \end{bmatrix}$$

A generalized unitary interactor matrix can be factored as

$$D_G = D_f D_{inf} = \begin{bmatrix} -0.6742q & -0.7385q \\ \frac{-0.7385(1-1.5477q)q}{q-1.5477} & \frac{0.6742(1-1.5477q)q}{q-1.5477} \end{bmatrix}$$

and the solution of the Diophantine identity yields

$$q^{-d}D_G N = q^{-1}D_G N$$

$$= \underbrace{\begin{bmatrix} -1.0434 & -0.3340 \\ \frac{0.6213-0.1340q^{-1}}{1-1.5477q^{-1}} & \frac{-1.7293+0.3731q^{-1}}{1-1.5477q^{-1}} \end{bmatrix}}_{F+q^{-d}R_{nmp}} +$$

$$+ q^{-1}\underbrace{\begin{bmatrix} \frac{-0.5217}{1-0.5q^{-1}} & \frac{-0.1670}{1-0.5q^{-1}} \\ \frac{0.0433}{1-0.5q^{-1}} & \frac{-0.1205}{1-0.5q^{-1}} \end{bmatrix}}_{R_{mp}}$$

150   12. Performance Assessment: User-specified Benchmark

So the closed-loop response under admissible minimum variance control is given by

$$q^{-d}D_G Y_t|_{admv} = (F + q^{-d}R_{nmp})a_t \quad (12.25)$$

Assume, for simplicity, that $Var(a_t) = I$, then the minimum variance achievable can be calculated from Equation 12.25 as

$$\Sigma_{admv} \triangleq Var(Y_t|_{admv}) = \begin{bmatrix} 1.6137 & -0.3521 \\ -0.3521 & 1.4988 \end{bmatrix} \quad (12.26)$$

and its quadratic measure (H_2 norm)

$$E(Y_t^T Y_t)|_{admv} = tr(\Sigma_{admv}) = 3.1125$$

Suppose that the controller sampling interval $\Delta T = 5$ s and a closed-loop response with time constant $\tau = 1$ min is desired. According to this requirement, a filter can be designed as

$$G_F = \begin{bmatrix} \frac{0.08}{1-0.92q^{-1}} & 0 \\ 0 & \frac{0.08}{1-0.92q^{-1}} \end{bmatrix}$$

Since $G_F(q^{-1} = 1) = I$, this specification preserves the asymptotic property of the type 1 system. From Equation 12.24, the user specified closed-loop response can be written as

$$q^{-d}D_G Y_t = (F + q^{-d}R_{nmp} + q^{-d}(I - G_F)R_{mp})a_t \quad (12.27)$$

Substituting numerical values into Equation 12.27 and simplifying the results gives

$$\begin{bmatrix} -0.67 + 0.96q^{-1} - 0.3q^{-2} & -0.74 + 1.05q^{-1} - 0.34q^{-2} \\ 1.14 - 2.36q^{-1} + 1.57q^{-2} - 0.34q^{-3} & -1 + 2.1q^{-1} - 1.38q^{-2} + 0.3q^{-3} \end{bmatrix} Y_t$$

$$= \begin{bmatrix} -1.5 + 1.96q^{-1} - 0.5q^{-2} & -0.5 + 0.6q^{-1} - 0.15q^{-2} \\ 0.66 - 1q^{-1} + 0.5q^{-2} - 0.06q^{-3} & -1.8 + 3q^{-1} - 1.5q^{-2} + 0.2q^{-3} \end{bmatrix} a_t$$

This is an achievable closed-loop response satisfying the user's requirement. Its variance can be calculated as

$$\Sigma_{user} \triangleq Var(Y_t|_{user}) = \begin{bmatrix} 2.3131 & 0.3130 \\ 0.3130 & 2.3248 \end{bmatrix}$$

and its quadratic measure (H_2 norm) as

$$E(Y_t^T Y_t)|_{user} = tr(\Sigma_{user}) = 4.6379$$

which is 50% larger than the achievable minimum variance. The performance indices are also adjusted accordingly. For example, the quadratic function based performance measure is adjusted according to

$$\eta_{user} = 1.5\eta_{admv}$$

where

$$\eta_{user} \triangleq \frac{tr(\Sigma_{user})}{tr(\Sigma_Y)}$$

12.5 Summary

Practical feedback control performance assessment has been discussed in this chapter. A filtered optimal H_2 control law with desired closed-loop dynamics has been proposed as a practical benchmark to assess control loop performance. The proposed approach has taken into account both minimum phase and nonminimum phase systems. Simulated examples have illustrated application of the proposed methods.

CHAPTER 13
PERFORMANCE ASSESSMENT: LQG BENCHMARK

13.1 Introduction

Many authors (Harris (1989), Desborough and Harris (1992), Stanfelj et al. (1993), Lynch and Dumont (1993), Kozub and Garcia (1993)) have reported the use of minimum variance control as a benchmark standard against which to assess control loop performance. This idea has been extended to MIMO processes in the previous chapters. However, these methods are concerned with performance assessment with minimum variance or optimal H_2 norm control as the benchmark, a benchmark which does not take into account the control effort explicitly.

In any case, minimum variance control is usually not the control algorithm of choice in most practical situations owing to its demand for excessive control action and poor robustness although performance assessment with minimum variance or optimal H_2 norm control as the benchmark does provide us with such useful information as a global lower bound of process variance or the H_2 norm measure. For example, if a controller indicates good performance relative to minimum variance control, then further tuning of the existing controller would be neither useful nor necessary. However, if a process indicates poor performance relative to minimum variance control, then there is potential to improve its performance but no guarantee that the performance will be improved by retuning the existing controller. In such cases, further analysis, such as performance evaluation with control action constraints taken into consideration, may be necessary. In general, tighter quality specifications result in smaller variation in the process output but typically require more control effort so one may be more interested in knowing how far away the control performance is from the "best" achievable performance with the same control effort, *i.e.*, in mathematical form the resolution of the following problem may be of interest:

Given $E[u_t^2] \leq \alpha$, what is $\min\{E[y_t^2]\}$?

The solution (achievable performance) is given by a tradeoff curve as shown in Figure 13.1. This curve can be obtained from solving the LQG problem (Kwakernaak and Sivan 1972; Harris 1985; Boyd and Barratt 1991), where the LQG objective function is defined by

$$J(\lambda) = E[y_t^2] + \lambda E[u_t^2]$$

By varying λ, various optimal solutions of $E[y_t^2]$ and $E[u_t^2]$ can be calculated. Thus, a curve with the optimal $E[u_t^2]$ as the abscissa and $E[y_t^2]$ as the ordinate is formed from these solutions. Boyd and Barratt (Boyd and Barratt 1991) have also shown that a variety of constraints such as hard constraints, robustness specification *etc.*, can be formed as convex optimization problems and are readily solved via convex optimization tools. Any linear controller can only operate in the region above the tradeoff curve (Boyd and Barratt 1991) shown in Figure 13.1. It is clear that given $E[u_t^2] = \alpha$, the minimum value (or the Pareto optimal value (Boyd and Barratt 1991)) of $E[y_t^2]$ can be found from this curve which represents the bound of performance, therefore, and can be used for performance assessment purpose.

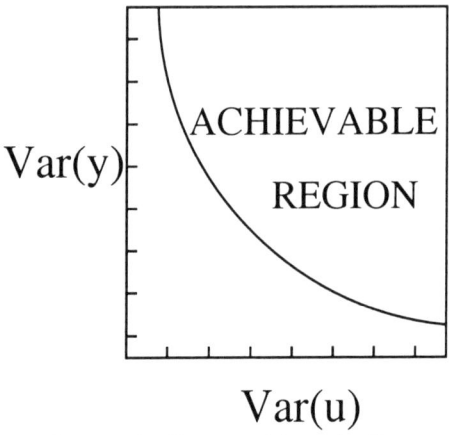

Fig. 13.1. An example of the tradeoff curve; this tradeoff curve separates achievable and non-achievable performance regions

13.2 Performance assessment with control action taken into account

13.2.1 LQG solution via state space or input-output model

The theory in this section is well known. Readers are referred to Kwakernaak and Sivan (1972) and Astrom and Wittenmark (1990). We will not discuss the theory but rather emphasize the LQG solution strategy via the Control Toolbox in MATLAB in this section.

Consider a state space model as

$$x_{t+1} = Ax_t + Bu_t + Gw_t$$
$$y_t = Cx_t + Du_t + v_t$$

13.2 Performance assessment with control action taken into account

where

$$E[w] = E[v] = 0, \quad E[ww^T] = Q, \quad E[vv^T] = R, \quad E[wv^T] = V$$

Then the Kalman filter can be written as

$$x_t^f = x_t^p + K^f(y_t - Cx_t^p - Du_t)$$

where x_t^p is the state prediction and can be written as

$$x_{t+1}^p = Ax_t^f + Bu_t$$

The steady state Kalman filter gain K^f can be solved via the Riccati Equation (Kwakernaak and Sivan 1972). The solution is readily available via a MATLAB function such as *dlqe*. If the state transition matrix A is singular, e.g., owing to time delays, then the function *dlqe2* in the MATLAB based MPC toolbox can be used. This uses the iterative method to solve the Riccati equation.

The optimal state feedback gain L ($u_t = -Lx_t^f$) is also solved via the Riccati Equation (Kwakernaak and Sivan 1972). The solution is readily available in the MATLAB function *dlqr*. For a singular matrix A, one could use *dare* in the LMI/MATLAB toolbox, which also uses an iterative method to solve the Riccati equation.

One can also solve the LQG problem with the input-output transfer function using the approach of Harris and MacGregor's (1987) which uses spectral factorization and the solution of the Diophantine identity.

13.2.2 LQG solution via GPC

Another way to solve this LQG problem is with the generalized predictive control (GPC) or model predictive control (MPC) approach (since the multivariate MPC/MATLAB toolbox is available). Consider a cost function of the form (Clarke, Mohtadi, and Tuffs 1987):

$$J_{GPC} = E\{\sum_{j=N_1}^{N_2} [y_{t+j} - r_{t+j}]^2 + \sum_{j=1}^{N_u} \lambda[\Delta u_{t+j-1}]^2\}$$

GPC gives a control law which "minimizes" the above objective function, but in order to achieve a time-invariant control law, only the first control action is actually implemented in GPC, *i.e.*, it is a receding horizon control law; therefore, the GPC control law does not truly minimize the above objective function. However, for $N_1 = 1$, $N_u = N_2$, and $N_2 \to \infty$, this objective function converges to the LQG objective function (Clarke, Mohtadi, and Tuffs 1987; Garcia, Prett, and Morari 1989; Bitmead, Gevers, and Wertz 1990), *i.e.*,

$$\frac{1}{N_2} J_{GPC} \to J_{LQG} = E[y_t - r_t]^2 + \lambda E[\Delta u_t]^2$$

As has been shown in Kwakernaak and Sivan (1972), the minimization of this LQG objective function yields a time-invariant optimal control law. Since the control law is time invariant for this special tuning, the GPC control law does optimize its objective function truly, irrespective of the fact that only the first control move is actually implemented; therefore, the LQG problem can be solved via the infinite GPC solution. But as $N_2 \to \infty$, GPC computation requires the solution of a large linear least squares problem, while LQG involves the solution of the recursive Riccati equation. Nevertheless, in practice, a finite value of N_2 is usually enough to achieve the approximate infinite horizon LQG solution via the GPC approach. Thus, the MPC toolbox in MATLAB provides a convenient approach to solve the LQG problem of MIMO processes.

13.2.3 The tradeoff curve

Once the problem is formulated as an LQG problem, the tradeoff curve can be calculated by varying the control weighting λ. For the SISO application, this is straightforward. To extend this result into MIMO systems, we need to explore this idea further. Suppose that the white noise sequence a_t satisfies $Var(a_t) = 1$. If $Var(a_t) = \sigma_a^2 \neq 1$, one can always normalize it to achieve such a form. For example, in the ARMAX form

$$Ay_t = Bu_t + Ca_t$$

If $Var(a_t) = \sigma_a^2 \neq 1$, then multiply the polynomial C such that $C' = C\sigma_a$, and the new ARMAX model can be written as

$$Ay_t = Bu_t + C'a'_t$$

where the new white noise sequence $a'_t = \sigma_a^{-1} a_t$ and, therefore, satisfies $Var(a'_t) = 1$. In the sequel, we assume, therefore, $Var(a_t) = 1$ without loss of generality.

Suppose that a regulatory LQG control law is $u_t = -\frac{E}{F} y_t$, then

$$y_t = \frac{CF}{AF + BE} a_t \overset{\triangle}{=} G_y a_t$$

and

$$u_t = -\frac{CE}{AF + BE} a_t \overset{\triangle}{=} G_u a_t$$

where F and E are functions of λ. The variance can be expressed as

$$Var(y_t) = Var[G_y a_t] = \frac{1}{2\pi} \int_{-\pi}^{\pi} |G_y(e^{j\omega})|^2 \sigma_a^2 d\omega = ||G_y||_2^2 \sigma_a^2 = ||G_y||_2^2$$

where the second equality holds by applying Parseval's theorem. Similarly,

$$Var(u_t) = ||G_u||_2^2 \sigma_a^2 = ||G_u||_2^2$$

13.2 Performance assessment with control action taken into account

Therefore, $Var(y_t)$ and $Var(u_t)$ are the H_2 norms of the closed-loop transfer functions from disturbance a_t to y_t and u_t respectively. Thus, the "size" of the closed-loop transfer function between y_t and a_t or between u_t and a_t is of primary concern in the performance assessment, irrespective of deterministic or stochastic nature of the disturbance a_t. In this way, the H_2-norm measure of y_t or u_t can also be applied to deterministic systems by replacing N to correspond to deterministic disturbances or a specific setpoint model. More importantly, this measure can be extended to the MIMO case directly, i.e., H_2 norms of the closed-loop transfer function matrices from disturbance a_t to Y_t and U_t are used for the performance measure.

For the MIMO case, the objective function of LQG control is written as

$$J = E[Y_t^T W Y_t] + \lambda E[U_t^T R U_t]$$

where the output weighting W should be chosen in such a way that it reflects the relative importance of the individual outputs; the control weighting R is also chosen according to the relative cost of individual control moves. By varying λ, various LQG control laws can be calculated. From the LQG control laws, one can calculate the closed-loop transfer function matrices from a_t to Y_t and U_t respectively as, for example, G_Y and G_U. Then the H_2 norm, $||G_Y||_W^2 = \{E(Y_t^T W Y_t)\}$ and $||G_U||_R^2 = \{E(U_t^T R U_t)\}$, can be used to plot the tradeoff curve. The procedure for constructing the tradeoff curve is summarized in Table 13.1

Table 13.1. The procedure for calculation of the LQG tradeoff curve

---

1. Formulate the problem in an appropriate LQG format.
2. Choose appropriate output weighting W and control weighting R.
3. By varying λ, a series of LQG control laws can be calculated. Using these LQG control laws, one can form the closed-loop transfer function from a_t to Y_t and U_t respectively.
4. The H_2 norms of the closed-loop transfer functions from a_t to Y_t and U_t respectively can then be calculated to provide a tradeoff curve.

13.2.4 Performance assessment

Using LQG as the benchmark to assess control loop performance requires a complete knowledge of the plant model, so an open- or closed-loop identification effort is required. Recent research on control relevant identification has shown that closed-loop identification is not necessarily poorer than open-loop identification if the objective of identification is control (Bialkowski 1993;

Gevers 1993; Van den Hof and Schrama 1995). For example, to design an LQG controller, the model is best identified under closed-loop conditions with the desired LQG controller running in the loop (Zang, Bitmead, and Gevers 1995). With an appropriate plant model, calculating the tradeoff curve is similar to designing a series of LQG controllers. In control-relevant identification, the "best" model is identified by using an iterative method, *i.e.*, identify a model; then re-design an LQG controller; re-identify the model using data collected under the LQG control; then re-design the LQG controller and so on. This approach is clearly not suitable for calculation of the tradeoff curve which requires a series of LQG designs. Identification of a "best" model for calculation of the LQG tradeoff curve remains an open problem for future research. In this section, the traditional identification methods are used to find the plant model instead. For example, we can identify the model under either closed-loop or open-loop conditions. Under closed-loop conditions, we can use direct identification or the two-step identification proposed in Chapter 15.

The noise model is also important for the solution of the LQG problem. It may be identified jointly with the plant model using routine operating data with dither excitation if regulation of these "routine" disturbances is of interest. The prediction error method (PEM) provides models of both plant and disturbances. On the other hand, one may want to assess control loop performance with hypothetical disturbances. For example, one may want to know how well a controller regulates step disturbances or tracks setpoints. In this case, the noise model (or setpoint dynamics) may simply be substituted by $1/(1 - q^{-1})$ if step-type disturbance regulation or tracking is of interest.

Once the tradeoff curve is calculated, the next step is to calculate the H_2-norm of Y_t and U_t under existing control in order to compare them with the tradeoff curve. Three different situations need to be considered.

If "routine" disturbances are of interest, then the white noise sequence a_t can be estimated from identification of the plant and noise model, *i.e.*,

$$\hat{a}_t = \hat{N}^{-1}(Y_t - \hat{T}U_t)$$

where $\hat{T}$ is the estimate of T and $\hat{N}$ is the estimate of N. With the estimate of the white noise sequence $\hat{a}_t$, the transfer functions from a_t to Y_t and U_t can be identified. Then the H_2 norms of Y_t and U_t are calculated.

If other hypothetical disturbances are of interest, then one needs to identify the sensitivity function S. The closed-loop transfer functions from a_t to Y_t and U_t can then be written as

$$Y_t = SNa_t$$

and

$$U_t = -SQNa_t$$

where

$$S = (I + TQ)^{-1}$$

If hypothetical setpoint changes (R_t) are of interest, then the tracking error E_t is considered. The closed-loop transfer functions from R_t to E_t and U_t can be identified and denoted as G_E and G_U. Then the H_2 norms of E_t and U_t are $||G_E N||_2^2$ and $||G_U N||_2^2$ respectively, where N characterizes the hypothetical setpoint dynamics.

13.3 Pilot-scale experimental evaluation

Example 13.3.1. To evaluate the proposed algorithm on a pilot-scale process, performance assessment of multivariate control loops was conducted on a two-interacting-tank process shown in Figure 13.2. The levels (h_1, h_2) of the two tanks are the two controlled variables. The two inlet flow rates (u_1, u_2) are the manipulated variables. The sampling interval is selected as $T_s = 5$ s. One step time delay (in addition to the delay due to the zero-order-hold) is introduced at the actuator of the control valve supplying water to Tank #2.

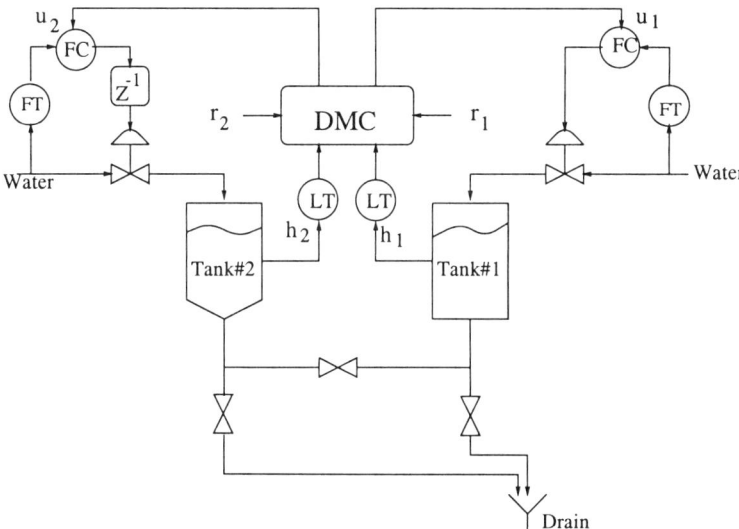

Fig. 13.2. Schematic diagram of the pilot-scale process.

IMC and GPC (or MPC) controllers were implemented on this process. To implement the IMC and GPC controllers, an open-loop identification test was conducted first to estimate the plant model. Using the *pem* function in MATLAB, the open-loop model was identified as

$$T = \begin{bmatrix} \frac{0.1963q^{-1}-0.1737q^{-2}-0.0112q^{-3}}{1-1.7208q^{-1}+0.7272q^{-2}} & \frac{0.0406q^{-7}-0.0113q^{-8}+0.0009q^{-9}}{1-0.6495q^{-1}+0.0482q^{-2}} \\ \frac{0.0147q^{-1}-0.0127q^{-2}+0.02q^{-3}}{1-1.3537q^{-1}+0.3707q^{-2}} & \frac{0.0406q^{-2}-0.0299q^{-3}-0.0047q^{-4}}{1-1.7849q^{-1}+0.7902q^{-2}} \end{bmatrix}$$

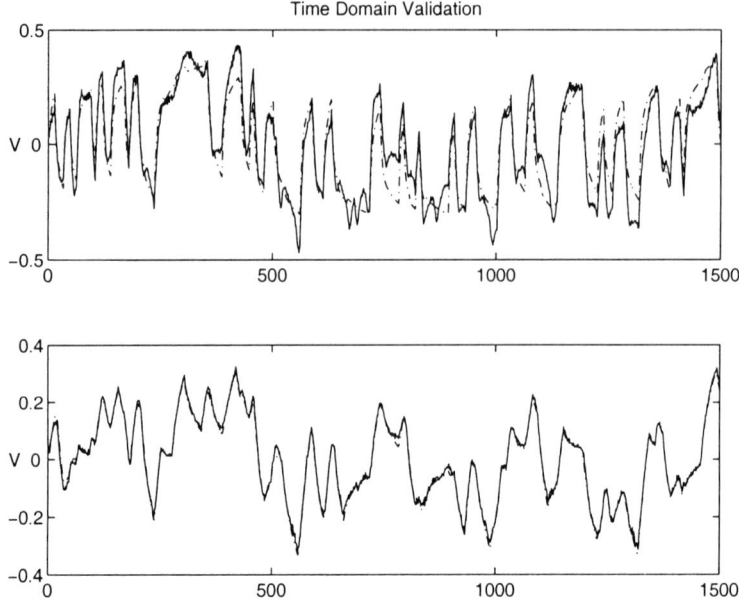

Fig. 13.3. Time domain validation of the open-loop model. The time scale is in terms of sampling intervals.

The time domain validation on a separate set of data is shown in Figure 13.3.
The hypothetical disturbance dynamics, in this example, are taken to be

$$N = \begin{bmatrix} \frac{1}{1-q^{-1}} & 0 \\ 0 & \frac{1}{1-q^{-1}} \end{bmatrix} \quad (13.1)$$

to represent step-type disturbances or setpoint changes. Based on the estimated plant model, three GPC controllers with different tuning parameters were implemented on this process. The tuning parameters are shown in Table 13.2. As $N_u = N_2$ and $N_2 \to \infty$, GPC converges theoretically to

Table 13.2. Controller tuning parameters

Controller	N_2	N_u	λ_1	λ_2
GPC #1	10	10	0.4	0.4
GPC #2	5	5	0.33	0.33
GPC #3	5	5	0.5	0.5

LQG solution. In this example, GPC converges to the infinite horizon case as $N_2 = N_u \geq 10$, so, theoretically, controller # 1 should give LQG performance. This may not be true owing to possible model-plant mismatch. We will evaluate the performance of these three controllers in this section.

A multivariate IMC controller was also implemented on this process. To design the IMC controller, a unitary interactor matrix has to be factored out from the plant transfer function matrix T. It is

$$D = \begin{bmatrix} -0.9972q & -0.0748q \\ 0.0748q^2 & -0.9972q^2 \end{bmatrix}$$

This unitary interactor matrix (an all-pass factor of the infinite zeros) represents time-delays in the MIMO system. The optimal IMC (Optimal H_2-norm) controller is the inverse of the delay-free transfer function matrix $\tilde{T}$, where

$$\tilde{T} = DT$$

To make the IMC controller implementable on this process, a filter

$$f = \begin{bmatrix} \frac{0.1}{1-0.9q^{-1}} & 0 \\ 0 & \frac{0.1}{1-0.9q^{-1}} \end{bmatrix}$$

has to be cascaded to the optimal IMC controller. The final IMC controller is

$$Q^* = \tilde{T}^{-1} f$$

where Q^* is the controller in the IMC framework (Morari and Zafiriou 1989). This IMC controller is termed "controller #4" in the following discussion.

Now we begin the evaluation of these four controllers using experimental data. Suppose we have no knowledge of the plant model and the controllers in this case. Closed-loop identification has to be used to identify this model in order to assess performance of the four controllers. A random binary dither signal is inserted in the setpoint. The closed-loop response (both Y_t and U_t of the four controllers are shown in Figures 13.4, 13.5, 13.6 and 13.7. One can get a rough indication of the relative performance of the four controllers by inspecting these four figures, but it is difficult to see how good these four controllers are relative to the best achievable control with the same control efforts.

A direct closed-loop identification using the prediction error method (Ljung 1998; Soderstrom and Stoica 1989) was applied to the four sets of data to estimate the plant model. Since the step-type setpoint tracking performance is of interest in this example, the setpoint dynamics require N to take the form expressed in Equation 13.1. The tradeoff curve was calculated and is shown in Figure 13.8. One can see differences in the curves from different sets of data. Clearly, these differences are not due to disturbance effects since the signal to noise ratio is fairly high in this experiment. It is mainly attributable to the bias error, *i.e.*, the model set is unlikely to contain the true dynamics of the plant. Thus, an "LQG relevant identification strategy" is highly desirable for such an application. The topic of control-relevant identification is beyond the scope of this book. Interested readers are referred to Kosut *et al.*

(1992), Shook et al. (1992), Bitmead (1993), Gevers (1993) and Van den Hof and Schrama (1995) for interesting discussions on this topic.

Since controller #1 is the closest to LQG control, in terms of performance, the tradeoff curve calculated from this data set is used as the benchmark in this example. By fitting the tracking error E_t and the controller output U_t to the setpoint dither excitation r_t directly, we can estimate the closed-loop transfer function matrices from r_t to E_t and U_t respectively. The fittings from r_t to E_t and from r_t to U_t are open-loop identification problems. Time domain validation (first 100 data points) for one of the experiments, controller #1, is shown in Figure 13.9. The upper panels represent time domain validation of the fitting from $r_t(2 \times 1)$ to $E_t(2 \times 1)$. The lower panels represent time domain validation of the fitting from $r_t(2 \times 1)$ to $U_t(2 \times 1)$. The other three sets of controllers show similar results and are not reproduced here. The H_2 norms of E_t and U_t, for the four sets of controllers, can be calculated using the setpoint dynamics as in Equation 13.1 and are shown in Figure 13.10. From this graph, one can compare the performance of different controllers or one controller with different tuning parameters. Of the three GPC controllers, controller #1 is the closest to LQG as it was expected to be. The difference between controller #1 and LQG may be attributed to model-plant mismatch. Controller #2 has the same "size" of tracking error as controller #1, but requires a larger control effort. Controller #3 exhibits a control effort which is similar to the IMC controller, controller #4, but yields a much smaller tracking error. We do not intend to show the superiority of any one controller over the other. In fact, any of the controllers can be retuned by adjusting the tuning or the filter parameters such that it moves toward the tradeoff curve. From this curve, one can see clearly the potential for improving the performance of controllers #2, #3 and #4. This study also provides insight into how the various GPC controllers may be tuned to give the desired performance.

13.4 Case study on an industrial process

Example 13.4.1. The proposed performance assessment method was applied to monitor control loop performance of the industrial cascade control loop shown in Figure 13.11. In this example, we are interested in control performance under a mode of regulation which rejects routine disturbances at the outer loop.

In Chapter 2. We have shown that the average performance index in the outer loop is approximately 0.15, indicating relatively poor control. Clearly this loop has the potential to provide better control by retuning the existing controller or redesigning the control algorithm, but one may ask how good this controller is relative to the achievable optimal control with the same control effort.

13.4 Case study on an industrial process

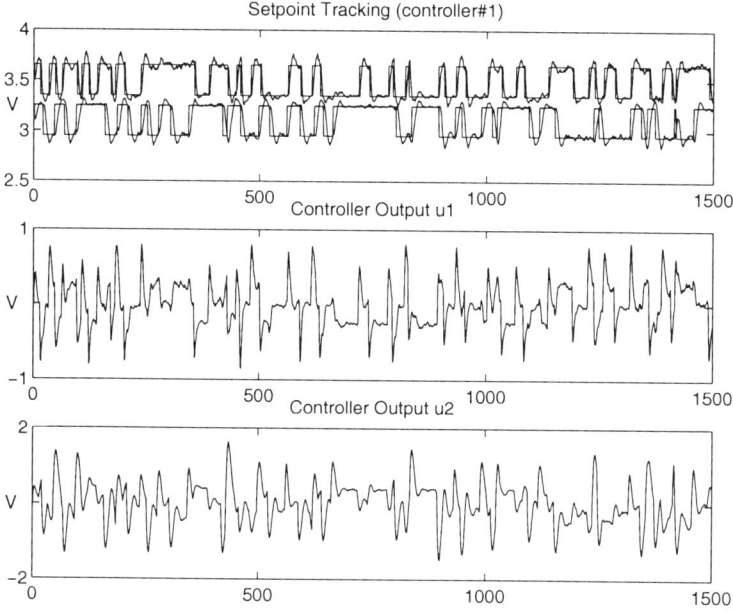

Fig. 13.4. Closed-loop test for controller #1. The time scale is in terms of sampling intervals.

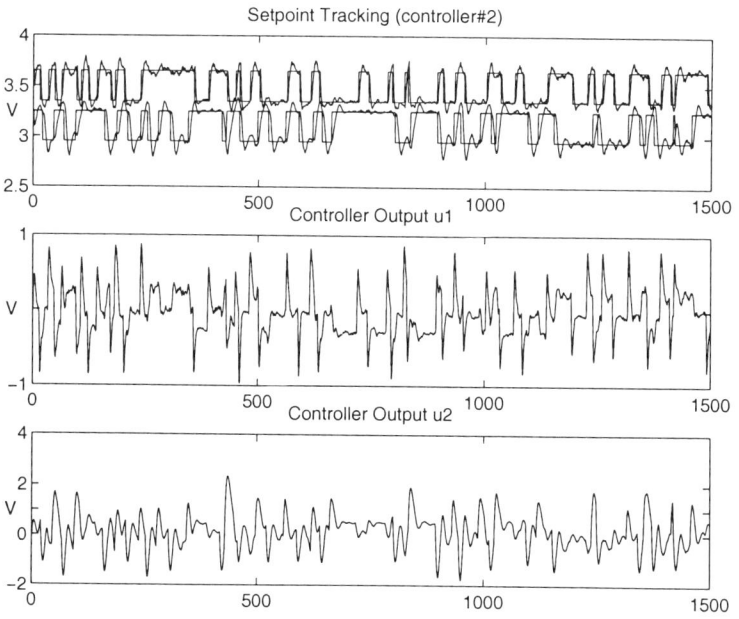

Fig. 13.5. Closed-loop test for controller #2. The time scale is in terms of sampling intervals.

164   13. Performance Assessment: LQG Benchmark

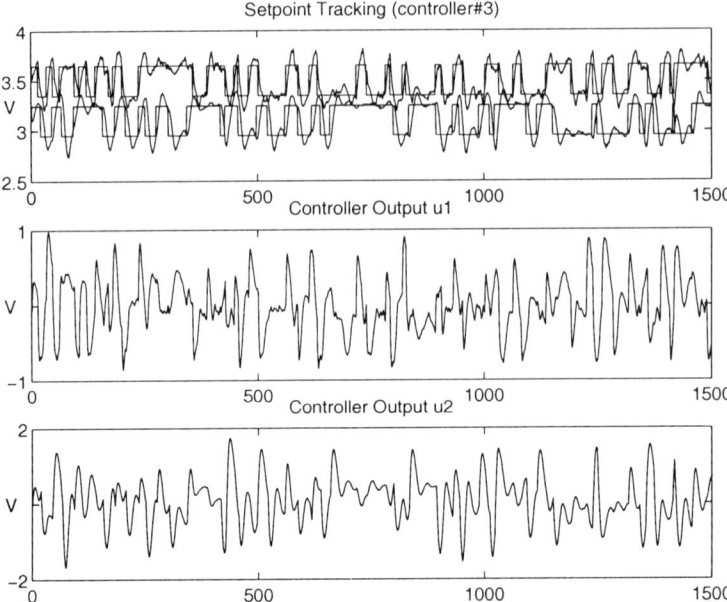

Fig. 13.6. Closed-loop test for controller #3. The time scale is in terms of sampling intervals.

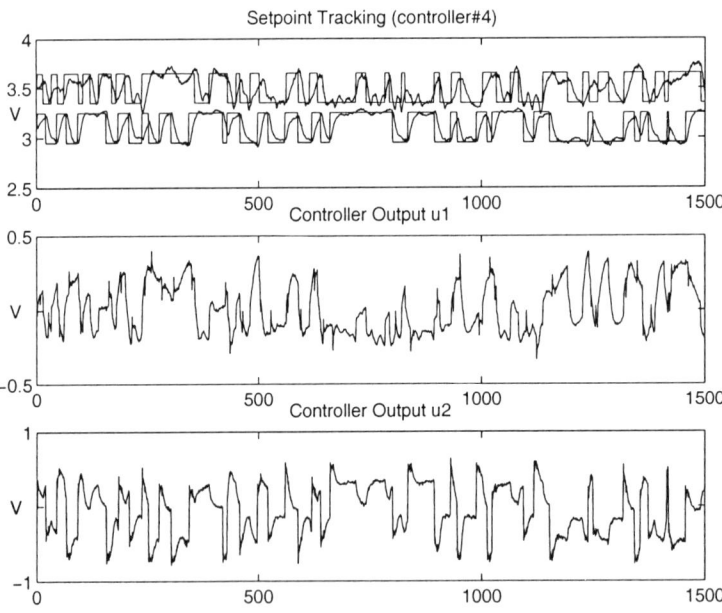

Fig. 13.7. Closed-loop test for controller #4. The time scale is in terms of sampling intervals.

13.4 Case study on an industrial process   165

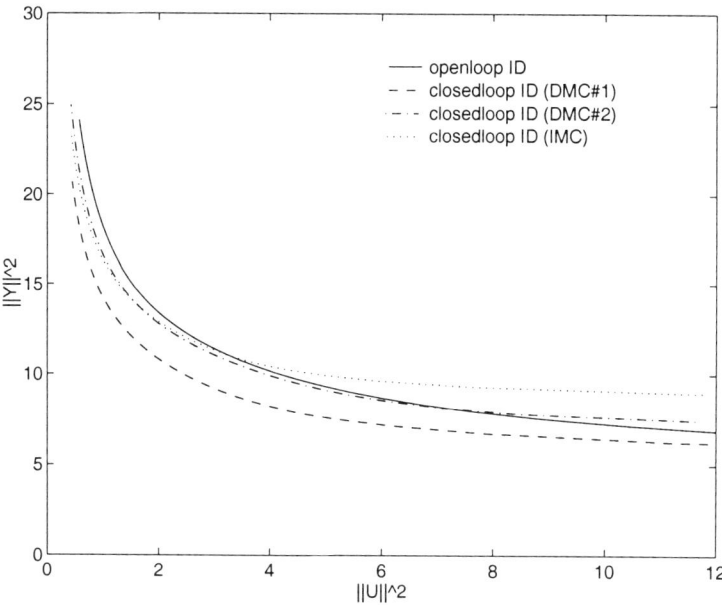

Fig. 13.8. Tradeoff curve estimated from different sets of data. Controller #3 (DMC#3) is not shown in the graph for clarity of the graph.

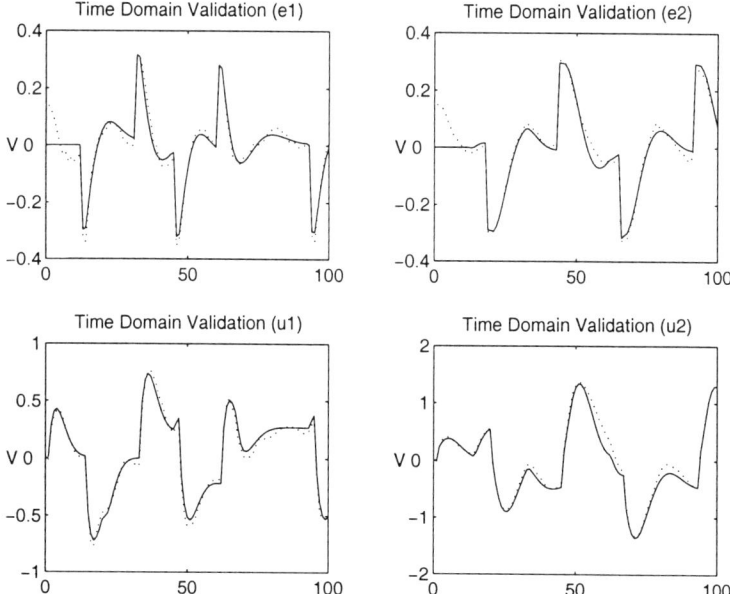

Fig. 13.9. Time domain validation for controller #1. The time scale is in terms of sampling intervals.

166   13. Performance Assessment: LQG Benchmark

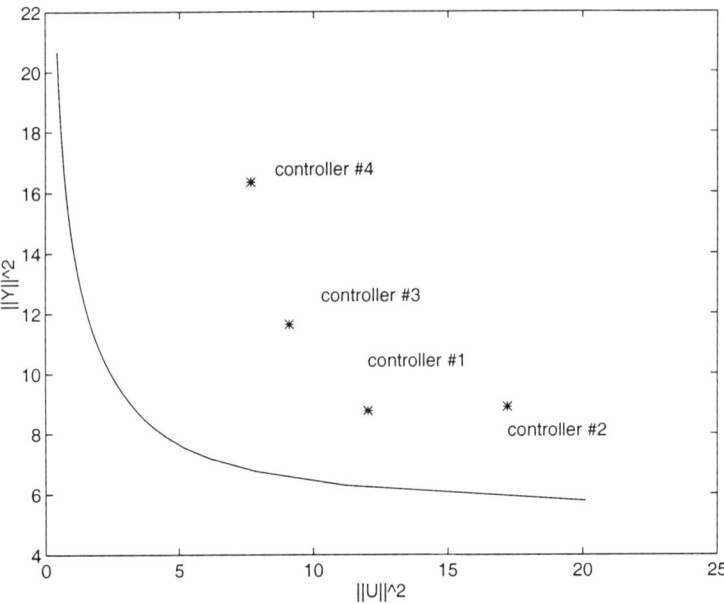

Fig. 13.10. Performance assessment of the four controllers.

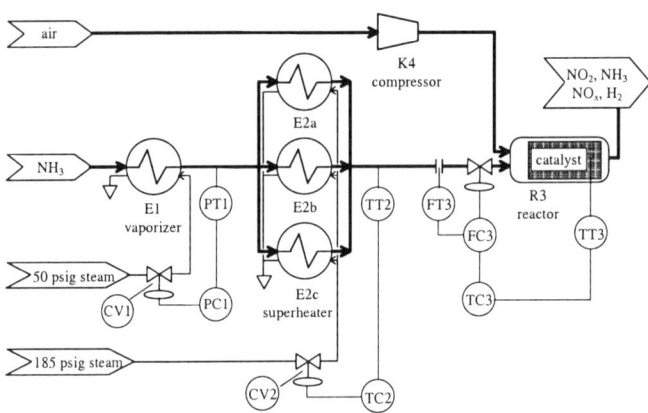

Fig. 13.11. Schematic diagram of the industrial cascade reactor control loop

An open-loop test was conducted on this process using PRBS excitation. Owing to the strong signal-noise-ratio in the experiment, a relatively good open-loop model is expected from open-loop data (Miller 1995), but the estimated noise model from open-loop data may not be completely reliable since the noise model may vary with the operating conditions. The routine closed-loop operation may be different from open-loop operation as well since a relatively large dither signal excitation was injected to perform the open-loop tests, *i.e.*, it may move the operation away from the nominal operating point. Consequently, the noise model obtained from open-loop tests may not represent the true noise dynamics under routine closed-loop conditions. For performance assessment purposes, a noise model which represents the noise dynamics under normal working conditions reasonably is necessary.

One approach to bypassing this problem is to use routine closed-loop operating data for the estimation of the noise model. This can be done by: 1) calculating the sensitivity function, *i.e.*, $S = 1/(1+TQ)$; 2) collecting routine closed-loop operating data and filtering the closed-loop data by the inverse sensitivity function, *i.e.*, $y^f = y/S$; 3) fitting the filtered closed-loop data by a time series model. This time series model is the estimated noise model from closed-loop data, and the procedure is based on the fact that routine closed-loop data y_t can be written as

$$y_t = \frac{N}{1+TQ} a_t$$

where N is the noise model and a_t is the white-noise excitation. If y_t is filtered by $1/S$, then $y_t^f = y(1+TQ) = Na_t$; therefore, N can be estimated from time series analysis of y_t^f. See Chapter 15 for further details.

With the plant model and the noise model, the tradeoff curve obtained from the LQG solution is shown in Figure 13.12. The abscissa represents the control variance measured by the expectation of incremental control action while the ordinate represents the variance of the temperature.

Using this graph, we can also assess the performance relative to minimum variance control. The graph indicates that the tradeoff curve converges (towards right) to 0.07 when there is no constraint on incremental control action Δu_t; therefore, the minimum variance (without incremental control action constraint) is $\sigma^2_{mv} = 0.07$. With the actual temperature variance $\sigma^2_y = 0.38$, the performance index $\eta(d) = \sigma^2_{mv}/\sigma^2_y = 0.18$.

The current variance of the incremental control action is about $E[\Delta u_t]^2 = 0.68$. Using this control action variance, the achievable temperature variance can be found from the curve which is about 0.1. Thus, the measure of achievable performance with LQG as the benchmark is $0.1/0.38 = 0.26$. This is a more realistic measure of current control performance when control action cannot be allowed to exceed the current level. This number indicates that there is significant potential to improve the feedback controller performance without increasing the control effort. If we draw a horizontal line along the actual working point, it intersects the tradeoff curve at the point where $E[\Delta u_t]^2 \to 0$. A significant reduction of the incremental control variance would be possible without increasing output variance if an advanced controller such as LQG or DMC were implemented.

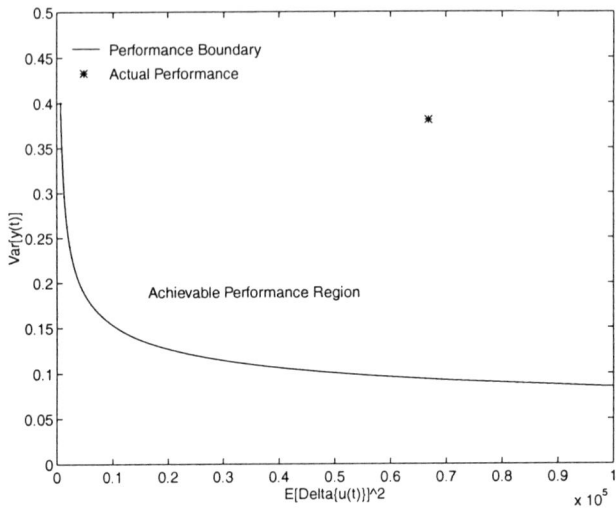

Fig. 13.12. Performance assessment of the industrial cascade control loop (outer loop)

13.5 Summary

A practical control benchmark has been proposed for the assessment of control loop performance. This practical benchmark takes both the control effort and the output performance into account. Calculation of such a benchmark control requires process identification. LQG is one such practical benchmark and its tradeoff curve can be obtained in terms of the H_2 norm of the appropriate transfer function matrices. Other practical benchmarks, which usually yield a similar tradeoff curve (Boyd and Barratt 1991) and require a numerical solution, may also be considered for further study of practical performance assessment.

CHAPTER 14
CONTROL LOOP PERFORMANCE MONITORING VIA MODEL VALIDATION

14.1 Introduction

Over the last twenty years, research on the detection of abrupt changes has emerged as an important area in the control community owing to the pioneering work of Basseville and Nikiforov (1993) and references therein. The problems concerned with this subject have focused, in the main, on fault detection and diagnosis, data segmentation, gain updating for adaptive algorithms, model validation and process quality control. Among various statistical-based detection algorithms, the local approach has recently regained significant interest owing to the notable work of Basseville (1998) and Zhang et al. (1998). The effectiveness and reliability of this approach has been demonstrated by its application in the monitoring of critical processes such as nuclear power plants, gas turbines, catalytic converter etc.(Basseville 1998). The local approach has a number of distinct features, among which are sensitivity to small changes and simplicity with the asymptotically uniformly most powerfulness. In addition, local approach algorithms can often be developed along the same lines as model identification. Thus, many established methods to enhance identification algorithms may be transferred readily to local detection algorithms.

In the previous chapters, we have discussed the use of minimum variance control as a benchmark standard against which to assess control loop performance. If variance reduction is the control objective, this technique is very appealing, and, in fact, its various algorithms have been applied in many industrial control loops, but minimum variance control has been found not to be a suitable benchmark for performance monitoring of model predictive control systems, where economic optimization is the main objective. The key to successful implementation of such control systems is, however, the process models. It is likely that poor performance of such control systems is due to poor models or changes in the model parameters. Thus, model validation and detection of model changes can be an effective way to assess and monitor performance of model predictive control systems. Industries are increasingly in need of a tool to validate their model continuously and to detect changes to the model-based control systems.

In this chapter, we will give an overview of the recent development in local detection algorithms that are capable of detecting small change while being robust enough to varying disturbance dynamics. Specifically, algorithms for monitoring parameter changes based on the ARX model and the output error model are discussed.

14.2 Problem formulation and the local detection approach

Consider a process shown in Figure 14.1 where T and N are a time-variant (open-loop or closed-loop) process transfer function and disturbance transfer function respectively. If T is an open-loop transfer function, then we are interested in model validation. If T represents a closed-loop transfer function, then the input to it should be a setpoint, *i.e.*, u_t is replaced by r_t. In this case, we are interested in control loop performance monitoring through the detection of changes in the closed-loop model parameters.

Let this time-variant process be described by one of the two dynamic models represented by parameters θ_0 and θ_1 respectively. The switching of the parameters occurs at some unknown time t_0:

$$y_t = Tu_t + Na_t \qquad (14.1)$$

where

$$\theta_0 = \begin{bmatrix} para(T_0) & para(N_0) \end{bmatrix} \quad \text{for } t < t_0 \qquad (14.2)$$

and

$$\theta_1 = \begin{bmatrix} para(T_1) & para(N_1) \end{bmatrix} \quad \text{for } t \geq t_0 \qquad (14.3)$$

where $para()$ denotes parameters of the corresponding transfer function and a_t is a white noise sequence.

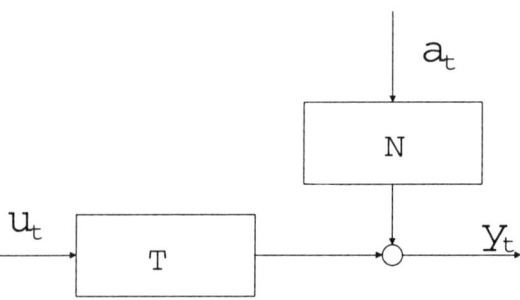

Fig. 14.1. Schematic of time-variant process

Significant change in the model parameters θ usually leads to significant change in the output behavior, and detection is relatively easy. It is the detection of the small change or early warning of process change that is more difficult and the most interesting to investigate in practice.

Now assume that parameters before change are given by θ_0, and after change are given by $\theta_0 + \frac{\eta}{\sqrt{N}}$ where η is a parameter vector with the same dimension as θ_0 but with an arbitrary direction and a small magnitude. By such an assumption of parameter change, we imply that: 1) the change is arbitrary and can happen in any one of or many of the parameters; 2) the change magnitude is small; and 3) with more data available, the detectability will increase, i.e., smaller changes can be detected by using a larger sample size.

With these assumptions, the change detection problem can be formulated as the following statistical test

$$H_0 : \theta = \theta_0 \quad \text{versus} \quad H_1 : \theta = \theta_0 + \frac{1}{\sqrt{N}}\eta$$

14.3 Normalized residuals and primary residuals

A statistic $\xi_N(\theta)$ given by

$$\xi_N(\theta) = \frac{1}{\sqrt{N}} \sum_{t=1}^{N} H(z_t, \theta)$$

is defined as a normalized residual if

$$E[H(z_t, \theta_0)] = 0 \qquad (14.4)$$

and

$$E[H(z_t, \theta)] \neq 0 \qquad \text{for } \theta \neq \theta_0 \qquad (14.5)$$

and $H(z_t, \theta)$ is defined as the primary residual(Basseville 1998), where $z_t = [u_t, y_t]^T$. Notice that the residuals defined here are different from those defined in most of the literature. The empirical version of Equations 14.4 and 14.5 may be written as

$$\frac{1}{N} \sum_{t=1}^{N} H(z_t, \theta_0) = 0 \qquad (14.6)$$

and

$$\frac{1}{N} \sum_{t=1}^{N} H(z_t, \theta) \neq 0 \qquad \text{for } \theta \neq \theta_0 \qquad (14.7)$$

If we can find such residuals, then for sufficiently large N, according to (Basseville 1998; Zhang, Basseville, and Benveniste 1998), we have

$$\xi_N(\theta_0) \sim N(0, \Sigma(\theta_0)) \qquad \text{under } H_0$$
$$\xi_N(\theta_0) \sim N(-M(\theta_0)\eta, \Sigma(\theta_0)) \qquad \text{under } H_1$$

where

$$M(\theta_0) = E(\frac{\partial}{\partial \theta} H(z_t, \theta)|_{\theta=\theta_0})$$

$$\Sigma(\theta_0) = \sum_{t=-\infty}^{\infty} Cov(H(z_1, \theta_0), H(z_t, \theta_0))$$

In practice, $M(\theta_0)$ may be approximated by

$$M(\theta_0) \approx \frac{\partial}{\partial \theta}[\frac{1}{N} \sum_{t=1}^{N} H(z_n, \theta)]_{\theta=\theta_0}$$

and $\Sigma(\theta_0)$ may be approximated by

$$\Sigma(\theta_0) \approx \frac{1}{N} \sum_{t=1}^{N} H(z_t, \theta_0) H^T(z_t, \theta_0) +$$

$$+ \sum_{i=1}^{I} \frac{1}{N-i} \sum_{t=1}^{N-i} (H(z_t, \theta_0) H^T(z_{t+i}, \theta_0) +$$

$$+ H(z_{t+i}, \theta_0) H^T(z_t, \theta_0))$$

where the value I should be properly selected according to the correlation of the signals. Typically, one can increase the value of I gradually until the result converges.

With these results, detection of small changes in the parameter θ is asymptotically equivalent to the detection of changes in the mean of a Gaussian vector. The generalized likelihood ratio (GLR) test for detecting unknown changes in the mean of a Gaussian vector is a χ^2 test. It can be shown that the GLR test of H_1 against H_0 can be written as

$$\chi^2_{global} = \xi_N(\theta_0)^T \Sigma^{-1}(\theta_0) M(\theta_0)$$
$$\times (M^T(\theta_0) \Sigma^{-1}(\theta_0) M(\theta_0))^{-1} M^T(\theta_0) \Sigma^{-1}(\theta_0) \xi_N(\theta_0)$$

If $M(\theta_0)$ is a square matrix, then this test can be simplified to

$$\chi^2_{global} = \xi_N(\theta_0)^T \Sigma^{-1}(\theta_0) \xi_N(\theta_0)$$

where χ^2_{global} has a central χ^2 distribution under H_0, and a noncentral χ^2 distribution under H_1. The degree of freedom of χ^2_{global} is the dimension of θ. A threshold value χ^2_α can be found from a χ^2 table, where α is the false alarm rate specified by the users. If χ^2_{global} is found to be larger than the threshold value, then a change in the parameter is detected.

14.4 Derivation of primary and normalized residuals - an illustrative example

Consider that variables $y_1(t)$, $y_2(t)$, $\cdots$, $y_n(t)$ from a multivariate process follow a linear regression model

$$y_1(t) = \beta_1 y_2(t) + \beta_2 y_3(t) + \cdots + \beta_{n-1} y_n(t) + v(t) \tag{14.8}$$

Notice that a dynamic ARX model

$$y_t + a_1 y_{t-1} + \cdots + a_{n_a} y_{t-n_a}$$
$$= b_1 u_{t-1} + \cdots + b_{n_b} u_{t-n_b} + a_t \tag{14.9}$$

is a special case of the linear regression model 14.8.

Define a parameter vector

$$\theta = \begin{bmatrix} \beta_1 & \beta_2 & \cdots & \beta_{n-1} \end{bmatrix}^T$$

Then the linear regression model can be written as

$$y_1(t) = x_t^T \theta$$

where

$$x_t = \begin{bmatrix} y_2(t) & y_3(t) & \cdots & y_n(t) \end{bmatrix}^T$$

The initial parameters or parameters before change can be calculated from the regression analysis. Recall that the objective of regression analysis is to minimize the following function

$$\begin{aligned} J &= \sum_{t=1}^{N} (y_1(t) - \beta_1 y_2(t) - \beta_2 y_3(t) - \cdots - \beta_{n-1} y_n(t))^2 \\ &= \sum_{t=1}^{N} (y_1(t) - x_t^T \theta)^2 \end{aligned}$$

Let the initial estimate or the parameters before change be given by θ_0. Then θ_0 can be calculated by solving the following equation

$$\frac{\partial J}{\partial \theta} = -2 \sum_{t=1}^{N} x_t (y_1(t) - x_t^T \theta) = 0$$

Therefore θ_0 must satisfy the equation

$$\sum_{t=1}^{N} x_t (y_1(t) - x_t^T \theta_0) = 0$$

Note that

$$y_1(t) - x_t^T \theta_0 = e(t)$$

which is the (regular) residual of the linear regression analysis.
Define

$$\xi_N(\theta) = \frac{1}{\sqrt{N}} \sum_{t=1}^{N} x_t(y_1(t) - x_t^T\theta) = \frac{1}{\sqrt{N}} \sum_{t=1}^{N} x_t e(t) \qquad (14.10)$$

Since

$$\frac{1}{N} \sum_{t=1}^{N} x_t e(t) = 0 \qquad \text{if } \theta = \theta_0$$

and

$$\frac{1}{N} \sum_{t=1}^{N} x_t e(t) \neq 0 \qquad \text{if } \theta \neq \theta_0$$

According to Equations 14.6 and 14.7, the primary residual is given by

$$H(z_t, \theta) = x_t e(t)$$

and $\xi_N(\theta)$ defined in Equation 14.10 is a normalized residual. Notice that the residuals $\xi_N(\theta)$ and $H(z_t, \theta)$ given here are vectors, while regular residual $e(t)$ by linear regression is a scaler and contains less information than the vector residuals.

Let

$$X = \begin{bmatrix} x_1^T \\ x_2^T \\ \vdots \\ x_N^T \end{bmatrix} \quad Y = \begin{bmatrix} y_1(1) \\ y_1(2) \\ \vdots \\ y_1(N) \end{bmatrix} \quad r = \begin{bmatrix} e(1) \\ e(2) \\ \vdots \\ e(N) \end{bmatrix} = Y - X\theta_0$$

Then

$$\xi_N(\theta) = \frac{1}{\sqrt{N}} X^T r$$

The gradient $M(\theta_0)$ is given by

$$\begin{aligned} M(\theta_0) &\approx \frac{\partial}{\partial \theta} [\frac{1}{N} \sum_{t=1}^{N} H(z_n, \theta)]_{\theta=\theta_0} \\ &= \frac{\partial}{\partial \theta} [\frac{1}{N} X^T (Y - X\theta)] \\ &= -\frac{1}{N} X^T X \end{aligned} \qquad (14.11)$$

14.5 Performance monitoring with varying disturbance dynamics

In many practical applications, processes are subject to numerous changes in disturbance dynamics. It may be in the interest of a monitoring scheme to detect the change of process model in the presence of changing disturbance dynamics. This problem is of particular interest in control loop performance monitoring, where performance deterioration may be simply due to a change in disturbance dynamics. It is possible that this change may be temporary and controller should not be re-tuned based simply on this transient behavior of the closed-loop dynamics. This problem is also of interest in model validation.

It is known(Ljung 1998; Soderstrom and Stoica 1989) that an ARX model can be used for an unbiased estimation of a dynamic model if the disturbance is white noise. For a process with correlated disturbances, a bias-free ARX model may be achieved by increasing the model order. An ARX model has an obvious advantage over other types of dynamic models in the sense that it can be estimated quickly using simple linear regression analysis. Thus, it may be utilized for quick and on-line monitoring although with the obvious drawback of its sensitivity to the choice of model order and change of disturbance dynamics. Since an ARX model is a special case of Model 14.8, the detection algorithm based on regression analysis, as discussed in Section 14.4, can be applied directly.

To improve the performance of the detection algorithm under the varying disturbance dynamics, the output error model(Ljung 1998) can be utilized since the output error identification algorithm does not parameterize disturbance models. Thus, changes of disturbance models do not issue alarms. Consider an output error model given by

$$y_t = \frac{B(q^{-1})}{F(q^{-1})} + v_t \tag{14.12}$$

where

$$B(q^{-1}) = b_1 q^{-1} + \cdots + b_{n_b} q^{-n_b}$$
$$F(q^{-1}) = f_1 q^{-1} + \cdots + f_{n_f} q^{-n_f}$$

and $v(t)$ is disturbance that is correlated and can be modeled by

$$v(t) = \frac{C(q^{-1})}{D(q^{-1})} a(t)$$

The identification of the output error model is to minimize the following objective function

$$J = \frac{1}{N} \sum_{t=1}^{N} \frac{1}{2} e^2(t, \theta) \tag{14.13}$$

where $e(t, \theta)$ is the prediction error given by

$$e(t,\theta) = y_t - \hat{y}(t|\theta)$$

and $\hat{y}(t|\theta)$ is a prediction given by

$$\hat{y}(t|\theta) = \frac{B(q^{-1})}{F(q^{-1})}$$

The gradient of Equation 14.13 can be calculated as

$$\frac{\partial J}{\partial \theta} = -\frac{1}{N}\sum_{t=1}^{N}\psi(t,\theta)e(t,\theta)$$

where $\psi(t,\theta)$ is the gradient of the prediction $\hat{y}(t|\theta)$ and can be calculated as

$$\psi(t,\theta) = \frac{\phi(t,\theta)}{F(q^{-1})}$$

where

$$\phi(t,\theta) = \begin{bmatrix} u(t-1) & \cdots & u(t-n_b) & -\hat{y}(t-1|\theta) & \cdots & -\hat{y}(t-n_f|\theta) \end{bmatrix}^T$$

Based on this result, following a similar procedure as in Section 14.4 one can derive the normalized residual as

$$\xi_N(\theta) = \frac{1}{\sqrt{N}}\sum_{t=1}^{N}\psi(t,\theta)e(t,\theta)$$

and the primary residual as

$$H(z_t,\theta) = \psi(t,\theta)e(t,\theta)$$

Example: Consider a closed-loop process given by

$$y_t = q^{-4}\frac{b}{1-fq^{-1}}r_t + \frac{1-cq^{-1}}{1-dq^{-1}}a_t \qquad (14.14)$$

with $\sigma_a = 0.6$. and initial (nominal operating) parameters $b = 0.33$, $c = 0.4$, $d = 0.67$ and $f = 0.67$. These nominal parameters may represent designed closed-loop dynamics, acceptable closed-loop dynamics, or representative closed-loop dynamics, subject to changes of both process and disturbance models. Results obtained from 100 Monte-Carlo simulation runs using ARX modeling are shown in Table 14.1 and 14.2 corresponding to 10% and 20% changes in parameters respectively. Results obtained from 100 Monte-Carlo simulation runs using OE modeling are shown in Table 14.3 and 14.4 corresponding to 10% and 20% changes in parameters respectively. The external perturbation (setpoint change) is a random binary signal with ±1 in the magnitude. The threshold is selected according to a false alarm rate of 1%. The 'detection rate' in the tables represents the number of successful detections of the process model change among the 100 simulation runs. The 'false rate' represents the false detection rate when, in fact, there is no change in the process model parameters (although with possible change of disturbance model parameters). All four tables show acceptable false alarm rates of near 1% if there is neither process change nor disturbance change. However, the OE based method has a much higher detection rate (power) than the ARX based method once there is a change in the process model. Furthermore, when there is no process model change, the false alarm rate in the presence of the changing disturbance model is reduced significantly by using the OE based method, indicating the robustness of the OE based approach to varying disturbance dynamics.

Table 14.1. Summary of Monte-Carlo detection simulation for 10% parameter change using ARX modeling

Process		Disturbance		Source	Detect.	False
b	f	c	d	of change	(%)	(%)
0.33	0.67	0.4	0.67	no change	N/A	2
0.37	0.67	0.4	0.67	process	60	N/A
0.33	0.74	0.4	0.67	process	79	N/A
0.33	0.67	0.44	0.67	disturbance	N/A	3
0.33	0.67	0.4	0.74	disturbance	N/A	18

Table 14.2. Summary of Monte-Carlo detection simulation for 20% parameter change using ARX modeling

Process		Disturbance		Source	Detect.	False
b	f	c	d	of change	(%)	(%)
0.33	0.67	0.4	0.67	no change	N/A	2
0.40	0.67	0.4	0.67	process	100	N/A
0.33	0.80	0.4	0.67	process	100	N/A
0.33	0.67	0.48	0.67	disturbance	N/A	8
0.33	0.67	0.4	0.80	disturbance	N/A	22

Table 14.3. Summary of Monte-Carlo detection simulation for 10% parameter change using OE modeling

Process		Disturbance		Source	Detect.	False
b	f	c	d	of change	(%)	(%)
0.33	0.67	0.4	0.67	no change	N/A	0
0.37	0.67	0.4	0.67	process	81	N/A
0.33	0.74	0.4	0.67	process	99	N/A
0.33	0.67	0.44	0.67	disturbance	N/A	0
0.33	0.67	0.4	0.74	disturbance	N/A	1

Table 14.4. Summary of Monte-Carlo detection simulation for 20% parameter change using OE modeling

Process		Disturbance		Source	Detect.	False
b	f	c	d	of change	(%)	(%)
0.33	0.67	0.4	0.67	no change	N/A	0
0.40	0.67	0.4	0.67	process	100	N/A
0.33	0.80	0.4	0.67	process	100	N/A
0.33	0.67	0.48	0.67	disturbance	N/A	0
0.33	0.67	0.4	0.80	disturbance	N/A	1

14.6 Summary

The method for detection of abrupt changes for process monitoring and control loop performance monitoring has been discussed in this chapter. The local detection approach with ARX modeling and OE modeling has been proposed and tested both theoretically and through simulation. The advantage of the proposed local approach based on OE modeling is its effectiveness in detecting small changes and its robustness to varying disturbance dynamics.

CHAPTER 15
CLOSED-LOOP IDENTIFICATION

15.1 Introduction

A necessary prerequisite for model-based control is a model of the process. Such certainty-equivalence, model-based control schemes rely on an off-line estimated model of the process, *i.e.*, the process is "probed" or excited by a carefully designed input-signal under open-loop conditions and the input-output data are used to generate a suitable model of the process. In a majority of model-based control schemes used in the chemical process industry, the models are generated with little regard for their ultimate end-use, e.g., as in model-predictive control. Almost always, in such cases, reduced-complexity models are generated to capture the most dominant dynamics of the process. Such batch or off-line identification methods represent a major effort and may require anywhere from several hours to several weeks of open-loop tests.

In contrast with this, the objective in closed-loop identification is to use routine operating data with dither signal excitation to develop a dynamic model of the process. Practically, it is a very appealing idea. In this mode, process identification can commence with the process in its natural closed-loop state. In some cases, the plant has to run under closed-loop conditions owing to safety reasons. In other cases, if a linearized dynamic model around a nominal operating point is desired, this can be achieved by closed-loop identification. Otherwise, under open-loop conditions, the process variables may drift away from the nominal operating point.

Closed-loop identification has been an area of considerable academic and industrial interest over the last 20 years. Important issues such as identifiability under closed-loop conditions have received attention from many researchers (Box and MacGregor 1976; Goodwin and Payne 1977; Gustavsson, Ljung, and Soderstrom 1978; Soderstrom and Stoica 1989; Forssell and Ljung 1999). A number of identification strategies have been developed (Ljung 1998; Soderstrom and Stoica 1989). In traditional identification literature, the quality of identification and identifiability issues are addressed mainly under the assumption that the model set contains the plant *i.e.*, the model can describe true process dynamics.

The more typical case is that of under-modelling or identification of reduced-complexity models when the plant is not in the model set. This is the focus of the present chapter and is a more realistic situation since a plant is generally of relatively high-order, and the model structure used to approximate such a process is almost always lower order. Stated simply, identification is an exercise in model-reduction. Under such circumstances, the model-plant error consists of two terms: the bias error which is due to under-modelling and the random error (or variance error) which is due to noise and disturbance effects. Several general expressions for the asymptotic variance and bias errors have been given by Ljung (1998) . The relationship between the variance and bias errors has been addressed recently by Guo and Ljung (1994) and Ljung (1994). These general expressions for the bias distribution have also been extended to closed-loop identification (Bitmead, Gevers, and Wertz 1990; Zhu and Backe 1993; MacGregor and Fogal 1995). It is well known that a data pre-filter can change the distribution of the bias error in the frequency range of interest (Ljung 1998), and therefore plays a somewhat similar role to the change in the frequency content of the dither signal. The choice of this pre-filter or, alternatively, the spectrum of the input signal is application dependent (Gevers and Ljung 1986; Ljung 1998; Rivera, Pollard, and Garcia 1992).

Closed-loop identification has also attracted much interest owing to the emerging research area of joint identification and control. The key idea in the joint identification and control strategy (as opposed to a 'disjoint' or separate identification and control) is to identify *and* control with the objective of minimizing a joint global control performance criterion. This topic has received attention under such headings as: control-relevant identification, iterative identification and control *etc.* Readers are referred to Kosut *et al.* (1992), Shook *et al.* (1992), Bitmead (1993) , Gevers (1993) and Van den Hof and Schrama (1995) for detailed discussions on these topics. The study of control-relevant identification requires that the best identification strategy is to identify the process under feedback with the intended controller in use. For example, a model intended for the design of minimum variance control is best identified under minimum variance feedback control (Gevers and Ljung 1986), and similarly for LQG control (Zang, Bitmead, and Gevers 1995; Hakvoort, Schrama, and Van den Hof 1994), model reference control (Hjalmarsson, Gevers, Bruyne, and Leblond 1994) *etc.*

The purpose of this chapter, however, is to focus attention *only* on the identification of the process model under closed-loop conditions. The estimated model is shown to have asymptotically identical expressions for the bias and variance terms regardless of how the identification run is conducted, *i.e.*, irrespective of open-loop or closed-loop conditions. The estimated model can then be used subsequently for improving existing controller design, controller redesign, control loop performance assessment, general analysis, *etc.* For example, a model obtained from closed-loop data under PID control may be used for the design of a DMC controller. Control loop performance assessment techniques, as discussed in the previous chapters, do not require an explicit process model when minimum variance control is used as a benchmark, but if a more practical benchmark standard, such as LQG, is used for evaluating existing control loop performance, then more information about the process is required.

The main contribution of this chapter is the development of a two-step closed-loop identification algorithm which yields the same expressions for both variance and bias errors asymptotically as in open-loop identification. These results obviate the need to conduct expensive open-loop tests when simple closed-loop tests with dither signal excitation will suffice. This chapter illustrates the point that a suitable model of the process can be estimated from closed-loop data with appropriate filtering.

The chapter is organized as follows: Section 15.2 gives a comparison between open-loop and closed-loop identification in terms of variance and bias errors. In Section 15.3, a two-step closed-loop identification scheme is proposed, which yields the same expressions of variance and bias errors asymptotically as open-loop identification. The proposed algorithm is evaluated on simulated examples in Section 15.5, and a computer-interfaced pilot-scale plant in Section 15.6.

15.2 Accuracy aspects of closed-loop identification

Consider a linear SISO plant, illustrated schematically in Figure 15.1, and described by

$$y = Tx + Na$$

where a is a white noise sequence, x is the input to the process, T is the process transfer function, and N is the noise transfer function. Let a model of the form

$$\hat{y} = \hat{T}x + \hat{N}a$$

be used to approximate the process dynamics. The prediction error is defined as

$$\hat{a} = \frac{1}{\hat{N}}(y - \hat{T}x)$$

The commonly used objective function for parameter identification is to minimize the sum of squares of the prediction error:

$$V = \frac{1}{M}\sum_{t=1}^{M}\hat{a}^2(t)$$

This method is termed the prediction error method or PEM (Ljung 1998). The total error of the estimates can be attributed to variance and bias errors (Ljung 1998) and may be written conceptually as a sum of the variance error and the bias error (Ljung 1994):

$$V_T = V_V + V_B$$

15. Closed-loop Identification

In this section, we deal with both the bias and variance estimation errors and compare them between the open-loop and closed-loop conditions. We show that the issue of variance and bias error of the parameter estimates is common to both open- and close-loop identifications. The PEM is a general and efficient method for system identification. The variance of the PEM estimator reaches the Cramer-Rao lower bound asymptotically (Ljung 1998). This asymptotic variance expressed in the frequency domain has been given by Ljung (1998).

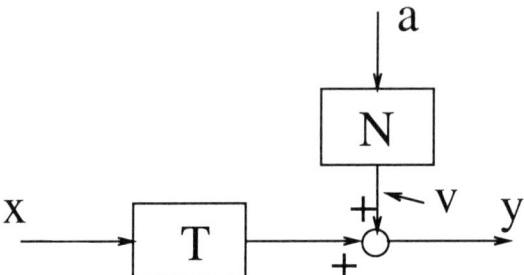

Fig. 15.1. Process model block diagram

Theorem 15.2.1. *(Ljung 1998) For input and output data x and y obtained from the process shown in Figure 15.1, where*

$$y = Tx + v = Tx + Na \qquad (15.1)$$

the following result holds for large sample size M, large model order n and small n/M:

$$Cov \begin{bmatrix} \hat{T}(e^{j\omega}) \\ \hat{N}(e^{j\omega}) \end{bmatrix} \sim \frac{n}{M} \Phi_v(\omega) \Phi^{-1}(\omega) \qquad (15.2)$$

where

$$\Phi(\omega) = \begin{bmatrix} \Phi_x(\omega) & \Phi_{xa}(\omega) \\ \Phi_{ax}(\omega) & \sigma_a^2 \end{bmatrix} \quad and \quad \Phi_v(\omega) = |N(e^{j\omega})|^2 \sigma_a^2$$

and where $\Phi_*(\omega)$ denotes the spectrum of the corresponding signal $*$.

Corollary 15.2.1. *For an open-loop system, with the input $x = w$ and output y where w is input excitation signal and is independent of noise a, Equation 15.1 can be written as*

$$y = Tw + v = Tw + Na \qquad (15.3)$$

The asymptotic variance of estimates using the PEM is given by

$$Var[\hat{T}(e^{j\omega})] = \frac{n}{M} \frac{|N(e^{j\omega})|^2 \sigma_a^2}{\Phi_w(\omega)} = \frac{n}{M} \frac{\Phi_v(\omega)}{\Phi_w(\omega)} \quad (15.4)$$

$$\frac{Var \hat{N}(e^{j\omega})}{|N(e^{j\omega})|^2} = \frac{n}{M} \quad (15.5)$$

Proof follows directly from Equation 15.2.

The results of Theorem 15.2.1 are more general than its statement may indicate. These results are also applicable to the direct closed-loop identification (Gevers and Ljung 1986). Direct closed-loop identification is identification of a plant model by directly using the input and output data regardless of the feedback effect since in this case the correlation between x and a, $\Phi_{xa}(\omega)$, is considered in the expression for $\Phi(\omega)$ (see Equation 15.2). In this way, Theorem 15.2.1 can also be extended to find the asymptotic variance of estimates $\hat{T}$ (Zhu and Backe 1993) and $\hat{N}$ under closed-loop conditions, i.e., $\Phi_{xa}(\omega) \neq 0$ (for the closed-loop case).

Corollary 15.2.2. *Under closed-loop control, as illustrated in Figure 15.2, the asymptotic variance of the estimates is given by*

$$Var[\hat{T}(e^{j\omega})] = \frac{n}{M} \frac{\Phi_v(\omega)}{\Phi_w(\omega)} \frac{1}{|S(e^{j\omega})|^2} \quad (15.6)$$

$$Var[\hat{N}(e^{j\omega})] = \frac{n}{M} |N(e^{j\omega})|^2 (1 + |Q(e^{j\omega})|^2 \frac{\Phi_v(\omega)}{\Phi_w(\omega)}) \quad (15.7)$$

where $S = 1/(1 + QT)$ is the sensitivity function; w is the dither signal and is independent of the noise sequence, a.

Proof. Under closed-loop conditions, the manipulated variable x can be written as

$$x = Sw - SQNa$$

Therefore,

$$\Phi_x(\omega) = |S(e^{j\omega})|^2 \Phi_w(\omega) + |S(e^{j\omega})|^2 |Q(e^{j\omega})|^2 |N(e^{j\omega})|^2 \sigma_a^2 \quad (15.8)$$
$$\Phi_{xa}(\omega) = -S(e^{j\omega})Q(e^{j\omega})N(e^{j\omega})\sigma_a^2 \quad (15.9)$$

The corollary follows after substituting Equation 15.8 and 15.9 into Equation 15.2.

Thus, the asymptotic variance of $\hat{T}$ under closed-loop conditions depends on the sample size M, the signal to noise ratio (SNR), $\Phi_w(\omega)/\Phi_v(\omega)$, and the sensitivity function $S(e^{j\omega})$. Increasing sample size tends to improve the estimate, but the sensitivity function affects the accuracy inversely. In process control, the asymptotic regulatory property under closed-loop control is of primary interest, i.e., it is desired to have asymptotic disturbance rejection or $\lim_{\omega \to 0} S(e^{j\omega}) = 0$. Typically, the disturbances are step-type, and therefore, such asymptotic disturbance rejection (for step-type disturbances) is achieved

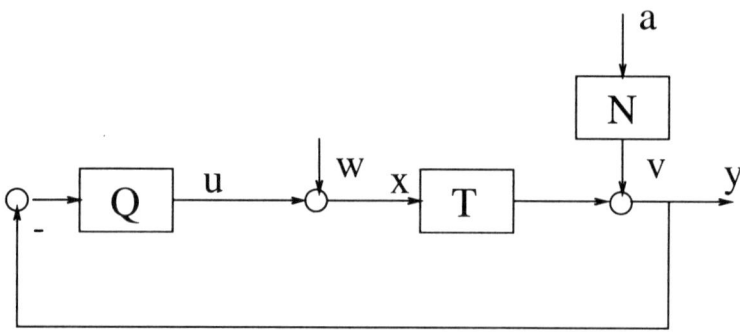

Fig. 15.2. Feedback control loop block diagram

by incorporating integral action. Consequently, the estimate in the lower frequency region is poor if a white noise dither signal is used, but one can also take advantage of the small value of the sensitivity at low frequency and use a dither signal, which has more power at low frequency, without upsetting the process. For other control strategies such as regulation of stochastic disturbances, the sensitivity function may have a small value in the middle or high frequency range, and, as a result, poor estimates in the middle or high frequency range would be expected. The main difference in the asymptotic variance between open-loop and closed-loop identification is the presence of the sensitivity function S (cf Equations 15.4 and 15.6).

In addition to the variance of the estimates, another important measure of identification quality is the bias error. It exists whenever the process dynamics are not contained in the model set, as in reduced-complexity model identification. In fact, this is almost always the case in practice. The distribution of bias errors in the frequency domain has been considered by Ljung (1998) through spectral characterization of the identification problem.

Theorem 15.2.2. *(Ljung 1998) For the open-loop process shown in Equation 15.3, the estimation of model parameters in the limit is given by the following optimization problem:*

$$\theta_M \stackrel{M \to \infty}{\Rightarrow} arg_\theta min \int_{-\pi}^{\pi} [|T(e^{j\omega}) - \hat{T}(e^{j\omega})|^2 \Phi_w(\omega) + \Phi_v(\omega)] \frac{1}{|\hat{N}(e^{j\omega})|^2} d\omega \quad (15.10)$$

where θ_M is the estimated model parameters. If the noise model $\hat{N}$ is chosen as being fixed, such as $\hat{N} = \bar{N}$, then the second term on the right hand side of Equation 15.10 is constant, and the optimization problem simplifies to

$$\theta_M \stackrel{M \to \infty}{\Rightarrow} arg_\theta min \int_{-\pi}^{\pi} |T(e^{j\omega}) - \hat{T}(e^{j\omega})|^2 \Phi_w(\omega) \frac{1}{|\bar{N}(e^{j\omega})|^2} d\omega \quad (15.11)$$

These results indicate clearly that the bias distribution of $|T(e^{j\omega}) - \hat{T}(e^{j\omega})|$ in the frequency domain is weighted by the dither spectrum, $\Phi_w(\omega)$, and the

15.2 Accuracy aspects of closed-loop identification   185

inverse of the noise spectrum, $1/|\hat{N}(e^{j\omega})|^2$ (also regarded as a noise filter), or simply the signal to noise ratio (SNR), $\Phi_w(\omega)/|\hat{N}(e^{j\omega})|^2$. If a unity noise model is considered, i.e., $\bar{N} = 1$, then the identification algorithm can be characterized as the output error method (OEM) (Ljung 1998) which does not depend on the noise spectrum:

$$\theta_M \stackrel{M\to\infty}{\Rightarrow} arg_\theta min \int_{-\pi}^{\pi} |T(e^{j\omega}) - \hat{T}(e^{j\omega})|^2 \Phi_w(\omega) d\omega \qquad (15.12)$$

Theorem 15.2.2 can also be extended to the closed-loop case.

Corollary 15.2.3. *(MacGregor and Fogal 1995) For the closed-loop process shown in Figure 15.2, the estimation of model parameters in the limit is given by the following optimization problem:*

$$\theta_M \stackrel{M\to\infty}{\Rightarrow} arg_\theta min \int_{-\pi}^{\pi} [|T(e^{j\omega}) - \hat{T}(e^{j\omega})|^2 \frac{|S(e^{j\omega})|^2}{|\hat{N}(e^{j\omega})|^2} \Phi_w(\omega) +$$

$$+ \frac{|S(e^{j\omega})|^2}{|\hat{S}(e^{j\omega})|^2} \frac{|N(e^{j\omega})|^2}{|\hat{N}(e^{j\omega})|^2} \sigma_a^2] d\omega \qquad (15.13)$$

where S is the sensitivity function and $\hat{S} = \frac{1}{1+Q\hat{T}}$.

Proof. See the proof in MacGregor and Fogal (1995).

If w is a sufficiently high order and persistently exciting signal, the set of $\hat{T}$ contains T, and the set of $\hat{N}$ contains N, then both estimates are consistent according to the direct identification method (Soderstrom and Stoica 1989), i.e., $\hat{T} \stackrel{M\to\infty}{\Rightarrow} T$ and $\hat{N} \stackrel{M\to\infty}{\Rightarrow} N$.

When the model set does not contain the process dynamics, which is generally the case and, as is of the interest in this chapter, a bias in estimation results, which is weighted once again, not only by the SNR but also by the sensitivity function S. Thus, the presence of the sensitivity function is the key difference between open-loop and closed-loop identification (cf. Table 15.1). To summarize, the expressions for the asymptotic variance and bias errors under open-loop and closed-loop conditions are listed in Table 15.1.

Table 15.1. Expressions of the asymptotic variance and bias errors

	open-loop	closed-loop
Var.	$\frac{n}{M} \frac{\Phi_v(\omega)}{\Phi_w(\omega)}$	$\frac{n}{M} \frac{\Phi_v(\omega)}{\Phi_w(\omega)} \frac{1}{|S(e^{j\omega})|^2}$
Bias distri.	$\int_{-\pi}^{\pi} [|T(e^{j\omega}) - \hat{T}(e^{j\omega})|^2 \frac{\Phi_w(\omega)}{|\hat{N}(e^{j\omega})|^2}$ $+ \frac{\Phi_v(\omega)}{|\hat{N}(e^{j\omega})|^2}]d\omega$	$\int_{-\pi}^{\pi} [|T(e^{j\omega}) - \hat{T}(e^{j\omega})|^2 \frac{|S(e^{j\omega})|^2}{|\hat{N}(e^{j\omega})|^2} \Phi_w(\omega)$ $+ \frac{|S(e^{j\omega})|^2}{|\hat{S}(e^{j\omega})|^2} \frac{|N(e^{j\omega})|^2}{|\hat{N}(e^{j\omega})|^2} \sigma_a^2]d\omega$

Remark 15.2.1. As pointed out by Schrama (1992), Gevers (1993) and Hjalmarsson et al. (1994), the "best" model for the joint identification and control design is not necessarily the "best" open-loop model. In fact, the "best" model for such design should have the bias error distribution weighted by the sensitivity function; however, this sensitivity function is precisely the sensitivity function that one wishes to "design" through the choice of a suitable model-based controller once a suitable model is available. In terms of identification and control, this represents a "catch-22" situation since an optimal controller cannot be designed if a control-compatible model is not available, and such a model cannot be estimated via closed-loop identification if its bias spectrum is not weighted by the appropriate sensitivity function. This is the main justification for iterative identification and control, but in the majority of model-based controller designs, the estimation will not have the appropriate sensitivity function as a weighting term. Throughout this chapter, we do not assume that the feedback controller under which the closed-loop data are collected is the intended or the ideal controller. In such cases, the dependence on the sensitivity function in place, based on existing control, should be decoupled in the first place. To circumvent this, one can decouple the effect of the current sensitivity function on the variance and bias errors, and also shape the bias distribution through the choice of appropriate data filters and the spectrum of the signal (for closed-loop identification) or the input signal (for open-loop identification). In this way, a model obtained via open or closed-loop identification can really serve the purpose of improved control law design, analysis or control loop performance assessment.

15.3 Two-step closed-loop identification

An effective way to reduce the variance of estimates is to increase sample size, but this may not have the desired effect on the reduction of the bias error. Depending on the application, smaller errors in some frequency ranges, e.g., around the cross-over frequency, may be desired while larger errors at other frequencies may be tolerated. Data prefiltering can change the distribution of the bias error over the frequency range of interest (Ljung 1998; Bitmead, Gevers, and Wertz 1990). MacGregor and Fogal(1995) have also shown that data prefilters and the noise model have significant effect on the bias error and identifiability for closed-loop identification. Under closed-loop conditions, the design of data filters is complicated by the presence of the sensitivity function. In this section, we propose a two-step closed-loop identification method which can decouple closed-loop parameter estimation from the effect of the undesired sensitivity function asymptotically. In doing so, this work provides a closed-loop identification method which retains the accuracy of open-loop identification asymptotically. Thus, many of the available open-loop experimental design techniques and data prefilters can be applied to closed-loop data. The most recent two-step identification algorithm proposed

15.3.1 Estimation of the sensitivity function—step 1

For a closed-loop system shown in Figure 15.2, the closed-loop response can be written as
$$y = TSw + NSa \qquad (15.14)$$
and
$$x = Sw - SQNa \qquad (15.15)$$

The sensitivity function, S, can be estimated from Equation 15.15, *i.e.*, Equation 15.15 presents a simple open-loop identification problem where the correlation between w and x yields $\hat{S}$. In order to apply Corollary 15.2.1 to analyze the variance error of the estimate, $\hat{S}$, the corresponding terms between Equation 15.3 and Equation 15.15 should be identified. The one-to-one correspondence between different terms in Equations 15.3 and 15.15 is summarized in Table 15.2.

Table 15.2. Item to item correspondence between two equations

Equation 15.3	Equation 15.15
y	x
w	w
T	S
N	$-SQN$
v	$-SQNa$

Using Table 15.2, the variance of the estimate of S can be found by applying Corollary 15.2.1 to Equation 15.15 as

$$\begin{aligned} Var[\hat{S}(e^{j\omega})] &= \frac{n}{M} \frac{\Phi_a(\omega)|N(e^{j\omega})|^2}{\Phi_w(\omega)} |S(e^{j\omega})|^2 |Q(e^{j\omega})|^2 \\ &= \frac{n}{M} \frac{\Phi_v(\omega)}{\Phi_w(\omega)} |S(e^{j\omega})|^2 |Q(e^{j\omega})|^2 \end{aligned}$$

and its relative variance as

$$\frac{Var[\hat{S}(e^{j\omega})]}{|S(e^{j\omega})|^2} = \frac{n}{M} \frac{\Phi_v(\omega)}{\Phi_w(\omega)} |Q(e^{j\omega})|^2 \qquad (15.16)$$

which depends on controller dynamics Q in addition to the sample size, the model order and SNR. In subsequent applications, we will show that only the relative accuracy of the sensitivity function is important.

Using Table 15.2, the bias distribution of the sensitivity function over the frequency range can be found by applying Theorem 15.2.2 to Equation 15.15. This yields the bias distribution in the frequency domain as

$$\theta_{SM} \xrightarrow{M \to \infty} arg_\theta min \int_{-\pi}^{\pi} [|S(e^{j\omega}) - \hat{S}(e^{j\omega})|^2 \Phi_w(\omega) + |H(e^{j\omega})|^2 \sigma_a^2] \frac{1}{|\hat{H}(e^{j\omega})|^2} d\omega \tag{15.17}$$

where $\hat{H}$ is the noise model with $H = -SQN$. A fairly large order can be selected since the sensitivity function serves as the first or intermediate result for subsequent identification of the process dynamics, i.e., the model set $\hat{S}$, should then be able to capture most of the dynamics of the actual sensitivity function, S. The total error (bias plus variance) would then be dominated by the variance error (Guo and Ljung 1994). The variance error is then the main issue of concern here. To achieve this, PEM may be used for the estimation of the sensitivity function, which has the property of asymptotic minimum variance. However, the distribution of the asymptotic minimum variance error over the frequency range cannot be controlled by the pre-filtering of the input-output data since both the process model and the noise model (or filter) are parameterized jointly in the PEM algorithm to yield asymptotic minimum variance estimates (Ljung 1998); therefore, the main "tuning knob" or "control parameter" for adjusting the relative variance of the sensitivity function (see Equation 15.16) is the spectrum of the dither signal, which has to be designed carefully in order to control the variance error over the frequency range of interest, but the variance error can also be reduced by increasing the number of data points. A relatively accurate estimate of the sensitivity function can be expected since it is not difficult to collect a relatively large number of data points under closed-loop operation. The bias error is assumed to be zero asymptotically by ensuring that the model set for the sensitivity function is large enough and includes the plant sensitivity function. For this reason we assume that $\hat{S} \to S$, in the following discussion.

15.3.2 Estimation of the process model—step 2

Once the sensitivity function is available, the process dynamics can be estimated by filtering output data with the inverse of the sensitivity function.

Proposition 15.3.1. *If one filters y by the inverse of the sensitivity function, and then applies PEM (joint parameterization of process and noise models) to Equation 15.14, the asymptotic variance of the estimates is determined by*

$$Var[\hat{T}(e^{j\omega})] = \frac{n}{M} \frac{\Phi_v(\omega)}{\Phi_w(\omega)} \tag{15.18}$$

and

$$\frac{Var[\hat{N}(e^{j\omega})]}{|N(e^{j\omega})|^2} = \frac{n}{M} \tag{15.19}$$

15.3 Two-step closed-loop identification

Thus, the variance of $\hat{T}$ and $\hat{N}/N$ is independent of closed-loop dynamics, i.e., both open-loop and closed-loop estimates have the same expression of accuracy with respect to variance (see Corollary 15.2.1).

Proof. Using $1/S$ to filter y yields the following relationship between y and w from Equation 15.14:

$$y/S = y^f = Tw + Na \qquad (15.20)$$

Identification of T from Equation 15.20 is an open-loop problem. This equation has the same form as Equation 15.3. Using w as input data and y^f as output data by applying Corollary 15.2.1, the proposition follows.

The bias distribution of the estimate over the frequency domain is also asymptotically independent of the closed-loop dynamics as shown in the following proposition:

Proposition 15.3.2. *From Equation 15.20, the asymptotic estimates of T and N from the filtered data y^f and w are given by the following optimization problem:*

$$\theta_M \stackrel{M\to\infty}{\to} \arg_\theta \min \int_{-\pi}^{\pi} [|T(e^{j\omega}) - \hat{T}(e^{j\omega})|^2 \Phi_w(\omega) + \Phi_v(\omega)] \frac{1}{|\hat{N}(e^{j\omega})|^2} d\omega \qquad (15.21)$$

Again, this yields the same bias distribution as under the open-loop condition (see Theorem 15.2.2). If both model sets, i.e., the process, $\hat{T}$, and the noise, $\hat{N}$, contain the true process dynamics, and w is a sufficiently high order and persistently exciting signal, then the parameter estimates as per Equation 15.21 will converge to the true values.

Proof. The proof follows by applying Theorem 15.2.2 to Equation 15.20.

Remark 15.3.1. In the proof of Proposition 15.3.1 and 15.3.2, it is assumed that the true sensitivity function S is used to filter y. If this sensitivity function is substituted by its estimate, $\hat{S}$, then Equation 15.20 should be written as

$$y/\hat{S} = y^f = \frac{S}{\hat{S}} Tw + \frac{S}{\hat{S}} Na$$

The validity of Propositions 15.3.1 and 15.3.2 will then depend on the relative accuracy of the sensitivity function, $S/\hat{S}$; therefore, the relative accuracy of the estimated sensitivity is of primary concern in the first step, but provided that $\hat{S}$ is sufficiently close to S (there is no model order or other structural limitation for the estimate of S), this relative error of the estimate of S should have a negligible effect on the bias expression in the estimates of T and N. On the other hand, the effect on the variance expression is not known exactly.

15. Closed-loop Identification

If only the process model T is to be estimated, then the output error method can be applied in this two-step identification approach. The estimation of both, the sensitivity function, S, and the process model, T, are open-loop identification problems. The consistency of the estimates $\hat{S}$ and $\hat{T}$ is independent of the noise model, as long as the noise model is fixed (Ljung 1998) as in the output error method.

Pre-filtering of data is important in identification. In particular, the choice of the data pre-filter can allow one to shape the spectral distribution or composition of the bias errors. The choice of the shaping filter should take into account the intended end-use of the model. This topic overlaps with the area of joint identification and control and has received much attention in the literature. The design and application of shaping filters for control loop performance assessment is a subject of considerable current interest in the literature. The point is that the sensitivity function decoupling filter provides a good starting point for the design of the shaping filter under closed-loop conditions. In the following proposition, we show that the bias error under the two-step identification strategy can be shaped freely.

Proposition 15.3.3. *(Shaping Filter) For the two-step identification, based on the output error method, if output data y is filtered by $F = G_f/S$, and the input data w is filtered by G_f only, where G_f is a shaping filter, then the asymptotic bias distribution is given by*

$$\theta_M \stackrel{M \to \infty}{\Rightarrow} arg_\theta min \int_{-\pi}^{\pi} [|T(e^{j\omega}) - \hat{T}(e^{j\omega})|^2 \Phi_w(\omega) |G_f(e^{j\omega})|^2] d\omega$$

Therefore, the bias distribution in the limit is independent of the closed-loop sensitivity function and can be shaped by the free filter G_f to meet the accuracy requirement over frequencies of interest.

Proof. Using filter $F = G_f/S$, the relationship between y and w from Equation 15.14 is now written as

$$y^f = Tw^f + G_f N a \tag{15.22}$$

where

$$y^f = \frac{G_f}{S} y$$
$$w^f = G_f w$$

Therefore,

$$\Phi_{w^f} = |G_f(e^{j\omega})|^2 \Phi_w(\omega) \tag{15.23}$$

By applying Theorem 15.2.2 with fixed noise model of unit value (*i.e.*, $\bar{N} = 1$), as in OEM, yields

$$\theta_M \stackrel{M \to \infty}{\Rightarrow} arg_\theta min \int_{-\pi}^{\pi} |T(e^{j\omega}) - \hat{T}(e^{j\omega})|^2 \Phi_{w^f}(\omega) d\omega \tag{15.24}$$

The proposition follows on substituting Equation 15.23 into 15.24.

The significance of this result is that the estimate obtained under closed-loop conditions can be shaped in the frequency domain if the model does not contain the true dynamics, while the estimator still maintains the property of consistency should the plant model ($\hat{T}$ only) contain the true dynamics. The classical closed-loop direct identification does not have such a property; therefore, all available methods for the design of the shaping filter for open-loop identification can be applied in this closed-loop case. For example, the open-loop, long-range predictive identification prefilter of Shook et al. (1992) and the systematic design of the control-relevant shaping filter of Rivera et al. (1992) can be applied. The choice of the shaping filter is analogous to selecting frequency weighting of the bias error function. Tighter weighting at some frequencies results in an expected corresponding reduction in bias errors at these frequencies but perhaps at the cost of larger bias errors at other frequencies. The effect of the shaping filter will be shown briefly in the experimental study of a pilot-scale process.

The aforementioned two-step identification algorithm is summarized in Tables 15.3 and 15.4.

Table 15.3. The procedure for two-step identification

1. Fit x to w by using the PEM or OEM, and obtain an estimate, $\hat{S}$, of the sensitivity function, S.
2. Filter y by $F = 1/\hat{S}$ and then fit y^f to w by applying the PEM. Then obtain estimates of T and N whose variance and bias expressions are asymptotically the same as open-loop identification.

Table 15.4. The procedure for two-step identification plus shaping

1. Fit x to w by using the PEM or OEM, and obtain an estimate, $\hat{S}$, of the sensitivity function S.
2. Filter y by $F = G_f/\hat{S}$ and w by G_f and then fit y^f to w^f by using the OEM. Obtain an estimate of T whose bias is asymptotically independent of the closed-loop sensitivity function and is shaped by the filter G_f.

Remark 15.3.2. If S contains non-minimum phase zeros, then $1/S$ cannot be used as an unstable decoupling filter. In this case, factorize S as

$$S = \frac{N^+ N^-}{D}$$

where the polynomial N^+ contains all non-minimum phase or unstable zeros. Let polynomial N^{+*} be a polynomial with its roots equal to the reciprocals of the non-minimum phase zeros in N^+. Then all roots of N^{+*} are inside the unit circle. Instead of using $1/S$ as the sensitivity function decoupling filter, $1/S'$ should be used as the decoupling filter to filter y_t, where $S' = (N^{+*}N^-)/D$. At the same time, w_t should be also filtered by N^+/N^{+*}. This will yield the same asymptotic properties (variance and bias) as using $1/S$ to filter y_t although, when S contains the unit-value zeros, the decoupling filter $1/S$ will have an integral term, which in some cases, may cause numerical problems. In this case, the probing or excitation signal is inserted preferably at the setpoint as discussed in the forthcoming sections.

Among many others (Caines and Chan 1975; Phadke and Wu 1974; Defalque, Gevers, and Installe 1976; Soderstrom and Stoica 1989), one of the most recent two-step identification strategies with a different objective has been proposed by Van den Hof and Schrama (1993) whose approach is summarized below.

Lemma 15.3.1. *Assume that the consistent estimate of the sensitivity function, $\hat{S} \rightarrow S$, is obtained in the first step. If the input data w is filtered by S, i.e., $w^f = Sw$, before applying OEM, then $\hat{T}$ can be estimated from the filtered data directly and is a consistent estimate.*

This is clearly seen from Equation 15.14, where

$$y = TSw + NSa = Tw^f + NSa \qquad (15.25)$$

whereas the approach proposed in this chapter considers filtering y by $1/S$ as follows:

$$y/S = Tw + Na$$

which is

$$y^f = Tw + Na$$

For brevity, the approach proposed by Van den Hof and Schrama is termed the w-filtering method, while the approach proposed in this chapter is termed the y-filtering method. In Van den Hof and Schrama (1993), "The sensitivity function is used to simulate a noise free input signal for an open loop identification of the plant to be identified. Using the output error method, an explicit approximation criterion can be formulated, characterizing the bias of identified models in the case of undermodelling".

15.3 Two-step closed-loop identification

Lemma 15.3.2. *By using the OEM, the w-filter approach yields the asymptotic frequency bias distribution as*

$$\theta_M \stackrel{M \to \infty}{\to} arg_\theta min \int_{-\pi}^{\pi} [|T(e^{j\omega}) - \hat{T}(e^{j\omega})|^2 |S(e^{j\omega})|^2 \Phi_w(\omega) d\omega$$

Thus, the frequency weighting on the bias $|T(e^{j\omega}) - \hat{T}(e^{j\omega})|^2$ *depends on the sensitivity function, S. In the approach proposed in this chapter, the frequency weighting on the bias is independent of the sensitivity function.*

This can be proved by applying Equation 15.12 in Theorem 15.2.2 to Equation 15.25. It should be pointed out that, under the framework of joint identification and control, the dependency of the bias error on the sensitivity function is not undesirable provided that the desired sensitivity function or the intended feedback controller is running during the data collection.

Remark 15.3.3. One of the main differences between the w-filtering approach and the y-filtering approach is whether w or y should be filtered by the sensitivity function or the inverse of the sensitivity function before carrying out the second step of identification. These two approaches result in different identification objectives. The y-filtering approach as proposed in this chapter aims at: 1) achieving the same "accuracy" expressions with respect to bias and variance errors under closed-loop and open-loop conditions (including consistency of the estimates if the model set contains the plant dynamics). This result is achieved by decoupling the closed-loop sensitivity function from closed-loop data; 2) obtaining explicit expressions for both asymptotic variance and bias errors. This approach is obtained by comparing the asymptotic variance and bias errors between open-loop and closed-loop conditions. The w-filtering approach as proposed by Van den Hof and Schrama provides: 1) a consistent estimate of the input-output transfer function if the model contains the plant dynamics; 2) an explicit expression for the asymptotic bias distribution only. The following illustrations show that these have important implications in closed-loop identification.

15.3.3 Other practical considerations

Until now our main consideration has been the case in which the dither signal is injected from w as shown in Figure 15.2. We will demonstrate that the general result can be extended to the case where the dither signal is injected from any point, for example via the setpoint r. Figure 15.3 shows an equivalent transformation of the block diagrams. The closed-loop response is now written as

$$y = STQr + SNa$$

This can be transformed to

$$\frac{1}{SQ}y = Tr + \frac{N}{Q}a$$

Therefore, if y is filtered by $1/SQ$ before applying the PEM or OEM, the relationship between r and y^f is

$$y^f = Tr + \frac{N}{Q}a$$

It is clear that both variance and bias distribution of estimates by using the PEM or OEM will be independent of the sensitivity function. Since it is again an open-loop identification problem, a shaping filter G_f can also be cascaded to the decoupling filter to shape the bias distribution in the frequency domain as illustrated in the foregoing discussion.

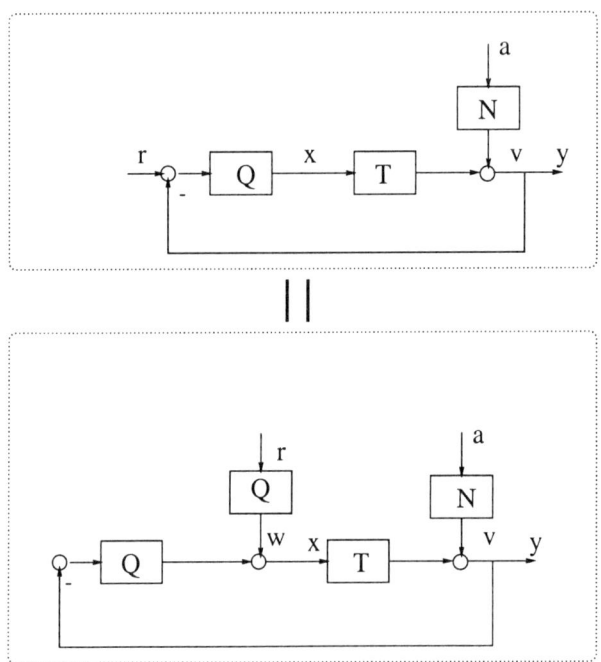

Fig. 15.3. Equivalent transformation of block diagrams.

In this case instead of estimating the sensitivity function S during the first step as in the foregoing section, QS should be estimated jointly. This estimate can be obtained by noticing the following relationship between r and x:

$$x = QSr - NQSa$$

Identification of QS using data r and x is an open-loop identification problem so it can be shown that the relative accuracy of the estimate of QS is independent of the sensitivity function. Since the inverse of QS does not contain the unit-value zeros introduced by integral control, it is the preferred sensitivity function decoupling filter when integral action exists in the controller.

15.4 Extension to MIMO systems

The two-step closed-loop identification can be extended to MIMO systems.

Proposition 15.4.1. *Under closed-loop conditions, the transfer function matrix T can be estimated via two steps. The sensitivity function is estimated from closed-loop data in the first step. The transfer matrix T from the sensitivity-filtered closed-loop data is then estimated in the second step.*

Proof. From Figure 15.2, we have

$$\begin{aligned} Y_t &= (I+TQ)^{-1}Tw_t + (I+TQ)^{-1}Na_t \\ &= STw_t + SNa_t \end{aligned} \quad (15.26)$$

where $S = (I+TQ)^{-1}$ and is termed as the sensitivity matrix. Filtering both sides of Equation 15.26 by the inverse sensitivity matrix $S^{-1} = (I+TQ)$ (or the return difference matrix (Bitmead, Gevers, and Wertz 1990)) gives

$$Y_t^f = S^{-1}Y_t = Tw_t + Na_t \quad (15.27)$$

This is clearly an open-loop identification problem. We also have

$$\begin{aligned} x_t &= (I+QT)^{-1}w_t - (I+QT)^{-1}QNa_{2t} \\ &= S_x w_t - S_x QN a_{2t} \end{aligned} \quad (15.28)$$

where $S_x \triangleq (I+QT)^{-1}$. The sensitivity function S can be written as

$$S = Q^{-1}(I+QT)^{-1}Q = Q^{-1}S_x Q \quad (15.29)$$

where the controller transfer function matrix Q either is known as *a priori* knowledge or can be identified from closed-loop data. Clearly, the estimation of S_x via Equation 15.28 is also an open-loop identification problem so the two-step identification can be achieved by: 1) estimation of the sensitivity function S via Equations 15.28 and 15.29; 2) identification of the transfer function matrix T via Equation 15.27.

15.5 Simulation

Example 15.5.1. Consider a second order ARMAX model with the transfer function given by

$$(1 - 0.79q^{-1} + 0.37q^{-2})y_t = (0.34 + 0.24q^{-1})u_{t-1} + (1 - 0.80q^{-1} + 0.12q^{-2})a(t)$$

A unity feedback control law is implemented in this simulation. The proposed y-filtering approach is compared with the direct identification method. The white noise a_t and the white-noise dither signal w_t are independent with $Var(a_t) = 2.25$ and $Var(w_t) = 1$ respectively. The number of data points in the simulation is $M = 5000$.

In general, identifiability under direct closed-loop identification requires that both the plant and disturbance dynamics lie in the set of plant and disturbance models (Soderstrom and Stoica 1989), but the w-filtering and y-filtering approaches do not have such a restriction in the choice of the noise model. The difficulty with the direct identification method is the choice (or tradeoff) of the plant model and the noise model, i.e., the plant model and the noise model are strongly coupled. One may choose high order models for both the plant and the noise, but this may violate the parsimony principle and also increase the variance error of the estimates as discussed in the previous sections; therefore, an incorrect choice of the noise model may yield an erroneous plant model and *vice versa*. In this example, we show that a first-order model, which is identified using the direct identification method and passes all residual tests, deviates significantly from the true dynamics. On the other hand, the y-filtering approach transforms the closed-loop identification to an open-loop identification problem and successfully detects the lack of fit when the first-order plant model is used.

Both the direct identification and y-filtering methods begin with a model of the first-order plant and second-order disturbances. There is clearly a model-order mismatch for such a choice of plant model. We will see which identification method can detect such a mismatch. Both methods use the Box-Jenkins model structure, i.e., BJ function in the System Identification toolbox in MATLAB. Residual tests for the models identified from both methods find the correlation between residuals and inputs, and, thus, indicate a lack of fit or a model-plant mismatch. This indicates that one may either increase the order of the noise model or increase the order of the plant model for the next trial.

To see the effect of the noise model, the noise models are increased to order three. The residual test for the direct identification is shown in Figure 15.4. The upper part of the figure shows the autocorrelation of the residuals and clearly indicates the "whiteness" of the residuals. The lower part of the figure is the cross correlation between residuals and past inputs, i.e., $E[\hat{a}_t u_{t-\tau}]/\sigma_{\hat{a}}\sigma_u$ for $\tau > 0$, where τ is the lag of the cross correlation function. Clearly, this cross correlation test indicates a sufficiently good fit of the data, i.e., no regions outside the 99% confidence intervals so the model obtained from direct identification passes the residual test, but the Bode diagram of the model shown in Figure 15.6 demonstrates a clear lack of fit. On the other hand, the residual test of the y-filtering identification is shown in Figure 15.5. The residuals also pass the "whiteness" test, but the cross correlation between the residuals and the inputs shows "spikes" outside the 99% confidence intervals and fails the test. Note that since the y-filtering approach transforms the closed-loop identification problem into the open-loop identification problem, the cross correlation test has to be carried out for both positive and negative legs (including the zero lag), i.e., the cross correlations between the residuals and all inputs (both past and future inputs) (Ljung 1998; Soderstrom and

Stoica 1989). Now we try to increase the order of the noise model further to a higher order, e.g., fifth order, for the y-filtering method while keeping the first-order plant model. The residual test is shown in Figure 15.7, and the model again fails the cross correlation test. This indicates that one has to increase the plant model order. Consequently, the plant model is increased to second order. The residual test is shown in Figure 15.8, and clearly, this model passes the residual test; therefore, the y-filtering method is able to find the correct model of the plant despite the mistake in the choice of the noise model. The Bode plot of the final estimate is shown in Figure 15.6.

The asymptotic variance of the estimate $Var(\hat{T})$ using the y-filtering approach is given in Equation 15.18. This equation is valid when the exact sensitivity function S is used as the decoupling filter, $1/S$, as shown in Proposition 15.3.1 and Remark 15.3.1. The predicted variance can be calculated from Equation 15.18 and is denoted by the dotted lines in Figure 15.9. To test validity of this predicted variance, 50 Monte-Carlo simulation runs are performed for this example. $\hat{T}$ is calculated from the two-step y-filtering approach using the exact sensitivity function as the decoupling filter. The variance of the estimate, $\hat{T}$, from 50 runs is calculated and also plotted in Figure 15.9 as the dash-dotted lines. Two cases with different data points for each simulation run are considered. The result for 512 data points is shown in the top part of the figure, and the result for 1024 data points is shown in the bottom part of the figure. The predicted 1σ bounds of the Nyquist plot are shown in Figure 15.10. Note that $Var(\hat{T})$ is defined by the variance of complex-valued random variables as

$$Var(\hat{T}) = E(\hat{T} - E\hat{T})(\hat{T} - E\hat{T})^*$$

(see (Ljung 1998)) where * means complex conjugate. From Figure 15.9, one can see that a good match in the low to medium frequency range is obtained in this example, but that the mismatch in the high frequency range is relatively large. The reasons could be: 1) the asymptotic variance as given by Ljung (1998) is an approximate expression, and should not be regarded as an exact expression; 2) the spectrum of dither and disturbances in each Monte-Carlo simulation run is different and may not be flat (white) over the frequency range of interest.

Remark 15.5.1. Direct closed-loop identification is simple and consistent if the model set contains the true plant and disturbance models. Whenever the model structure is known *a priori* or can be determined easily from data, then direct closed-loop identification is preferred since it gives a minimum variance estimate as well. However, this example also shows that direct closed-loop identification may give an erroneous result as well, owing to the bias error if the model structure is not chosen correctly.

Example 15.5.2. Consider an example in Schrama (1991). The plant under consideration consists of a transfer function, which is a discrete-time model of

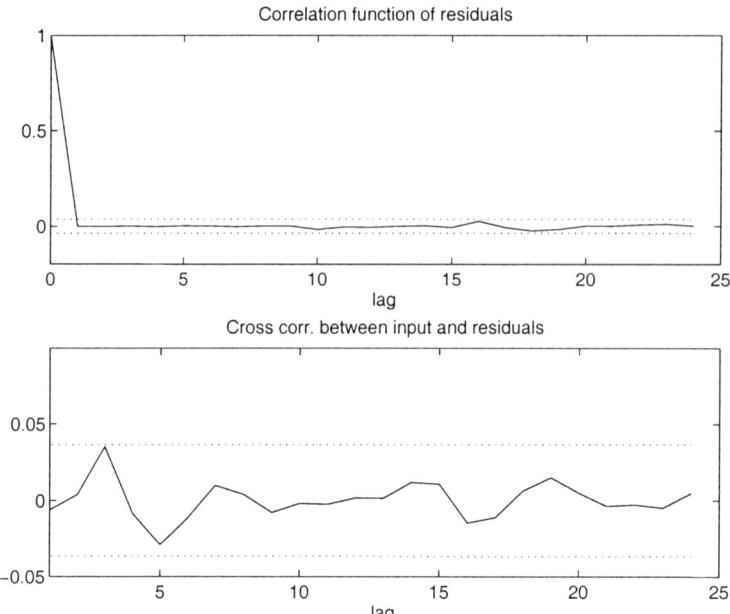

Fig. 15.4. Residual test for the model identified by using direct identification method (first-order plant and third-order noise).

a laboratory set-up, and some artificial noise contribution. The plant transfer function is given by

$$T = \frac{10^{-3}(0.98q^{-1} + 12.99q^{-2} + 18.59q^{-3} + 3.30q^{-4} - 0.02q^{-5})}{1 - 4.40q^{-1} + 8.09q^{-2} - 7.83q^{-3} + 4.00q^{-4} - 0.86q^{-5}}$$

According to Schrama (1991), in order to state a non-trivial case study, noise contributions are assumed to affect the input u and output y additively. The additive input noise is a white noise with variance $1/9$. The output noise is a white noise that is filtered by

$$N = \frac{0.01(2.89 + 11.13q^{-1} + 2.74q^{-2})}{1 - 2.70q^{-1} + 2.61q^{-2} - 0.90q^{-3}}$$

A control law

$$Q = \frac{0.61 - 2.03q^{-1} + 2.76q^{-2} - 1.83q^{-3} + 0.49q^{-4}}{1 - 2.65q^{-1} + 3.11q^{-2} - 1.75q^{-3} + 0.39q^{-4}}$$

is implemented in the plant. A dither signal with variance 1 is injected in the process in order to perform closed-loop identification. To demonstrate the effect of under-modelling, a fourth-order plant model (the original plant is fifth-order) is used for the identification.

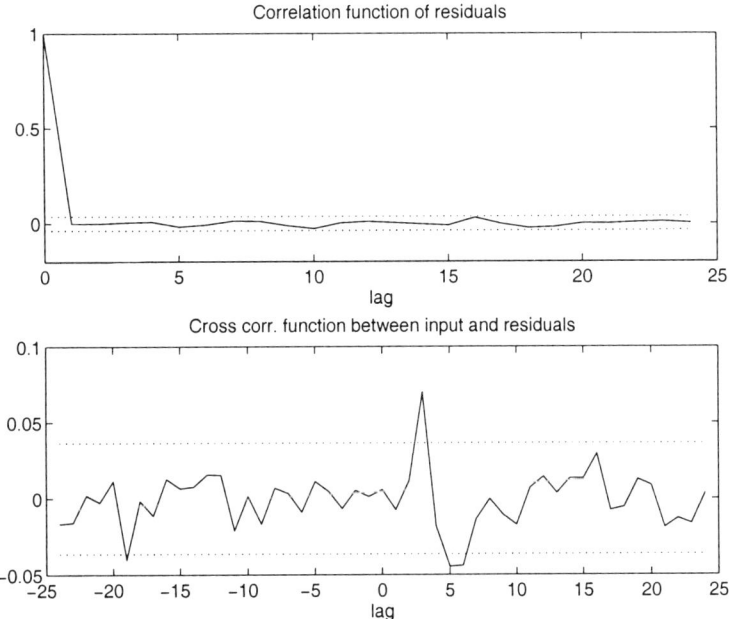

Fig. 15.5. Residual test for the model identified by using the y-filtering method (first-order plant and third-order noise).

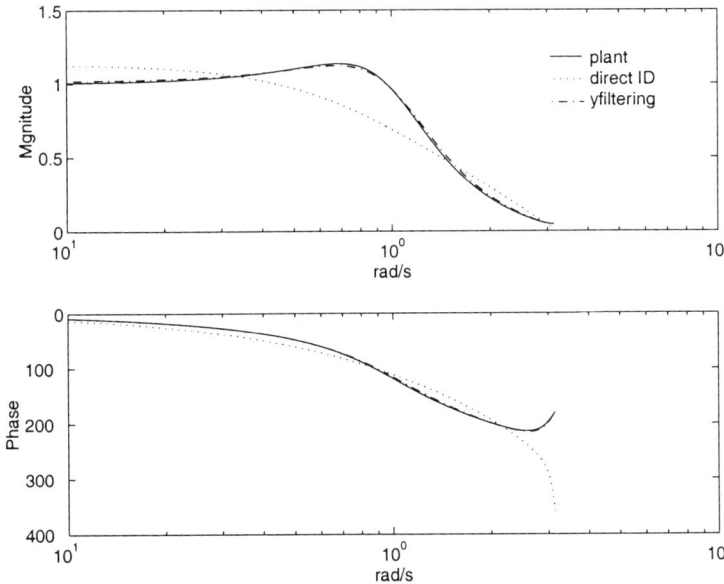

Fig. 15.6. Comparison between direct identification and the y-filtering methods.

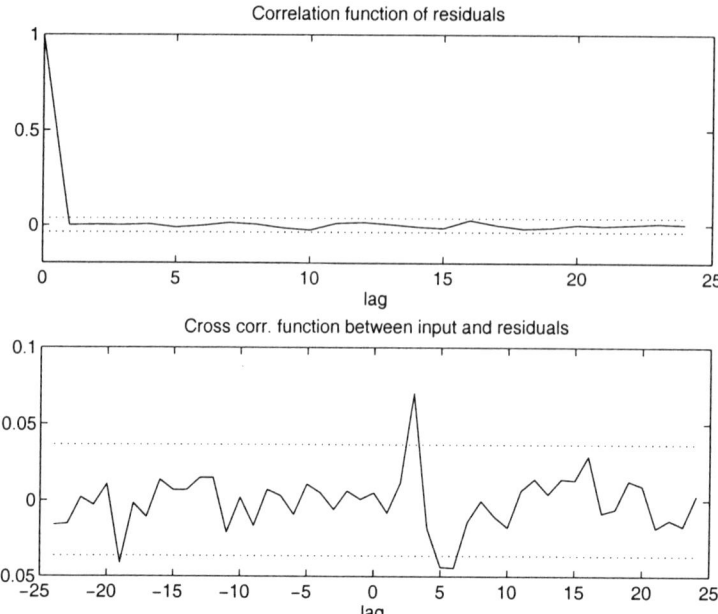

Fig. 15.7. Residual test for the model identified by using the y-filtering method (first-order plant and fifth-order noise).

The w-filtering approach and y-filtering approach are applied to the process and the results are presented in the Bode diagrams shown in Figure 15.11. Since the area of interest in this example is the plant model and the output error method is used for parameter estimation, the most relevant model validation is the cross-correlation test between residuals and inputs (Ljung 1998). The cross-correlation tests are performed with the results shown in Figure 15.12 and 15.13. Since both w-filtering and y-filtering transform the closed-loop identification to open-loop identification, the cross-correlation test should be conducted over the whole graph (*i.e.*, including both negative and positive legs). Clearly, the models obtained under w-filtering and y-filtering both pass the residual test.

Although both models have passed the time-domain test, the qualities of the models are significantly different in the frequency domain. If we look at the estimated sensitivity function shown in Figure 15.14, we can see the smaller magnitude of the sensitivity function in the medium frequency range with the minimum occurring around the frequency $\omega = 0.17 rad/s$. This shape of the sensitivity function is expected to affect the identification result, an expectation confirmed in Figure 15.11. The w-filtering approach gives a poor match in the medium frequency range including the cross-over frequency, particularly around the frequency $\omega = 0.17 rad/s$. The y-filtering, on the other hand, matches the true plant relatively well in the medium frequency

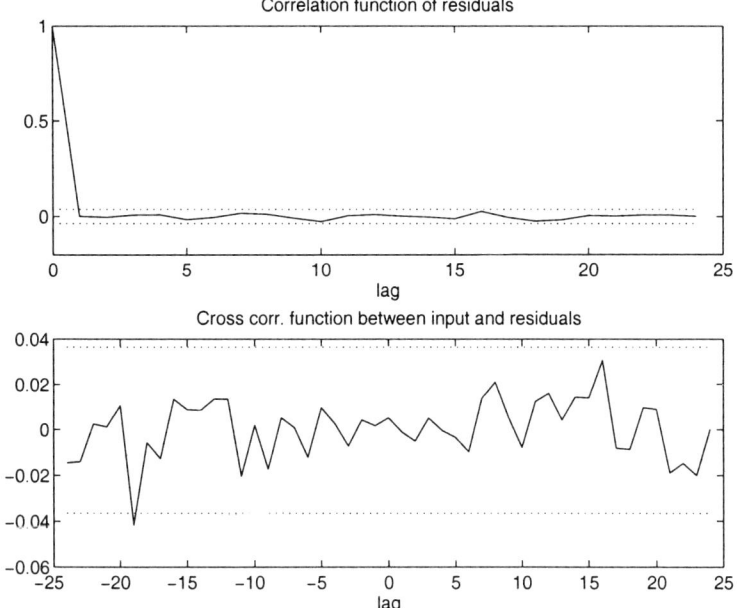

Fig. 15.8. Residual test for the model identified by using the y-filtering method (second-order plant and second-order noise).

range including the cross-over frequency even though this improvement is at the cost of the high frequency mismatch.

Example 15.5.3. The example used by Van den Hof and Schrama (1993) is considered in this Monte Carlo simulation for comparison of different approaches. The discrete plant (causal but not strictly causal) is represented by transfer functions

$$T = \frac{1}{1 - 1.6q^{-1} + 0.89q^{-2}}$$

$$Q = q^{-1} - 0.8q^{-2}$$

$$N = \frac{1 - 1.56q^{-1} + 1.045q^{-2} - 0.3338q^{-3}}{1 - 2.35q^{-1} + 2.09q^{-2} - 0.6675q^{-3}}$$

The noise signal a and the dither signal w are independent unit variance zero mean random signals. The number of data points for each run is chosen as $M = 2048$ in accordance with Van den Hof and Schrama (1993) and the simulation was run 50 times with different random seeds. Although this is an unrealistic plant (without any time-delay), mathematically this is a good simulation example to compare the sensitivity with the model structure mismatch for different identification schemes.

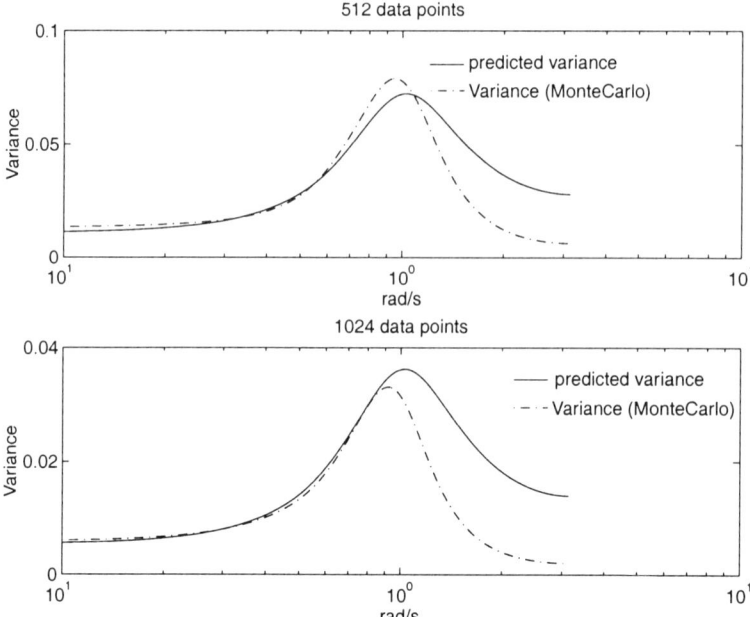

Fig. 15.9. Variance of the estimate calculated from Monte-Carlo simulation (second-order plant and second-order noise).

To compare sensitivity of the w-filtering, y-filtering and the direct PEM closed-loop identification with model-plant mismatch, one unit time-delay is considered in the model. Without model-plant mismatch, all of these three methods should give consistent estimates as discussed in the previous sections; therefore, a model of the following form is assumed to be

$$\hat{T} = \frac{(b_1 + b_2 q^{-1})q^{-1}}{1 + a_1 q^{-1} + a q^{-2}}$$

The estimated sensitivity function is shown in Figure 15.15. The sensitivity at lower frequency is smaller than at higher frequency. There is a valley with a minimum magnitude at the frequency of 0.18. This shape of the sensitivity function reduces the accuracy at lower frequency, and one would expect relatively large estimation errors around the frequency of 0.18 if the w-filtering approach is used (see Lemma 15.3.2). The lower portion of Figure 15.15 confirms this. Compared with the w-filtering approach, the y-filtering method clearly avoids the peak error with a slightly larger mismatch at high frequencies. The direct closed-loop identification does not work in this example owing to the structure mismatch. The comparison of the three approaches can also be clearly seen from the averaged Nyquist plot shown in Figure 15.16.

15.5 Simulation   203

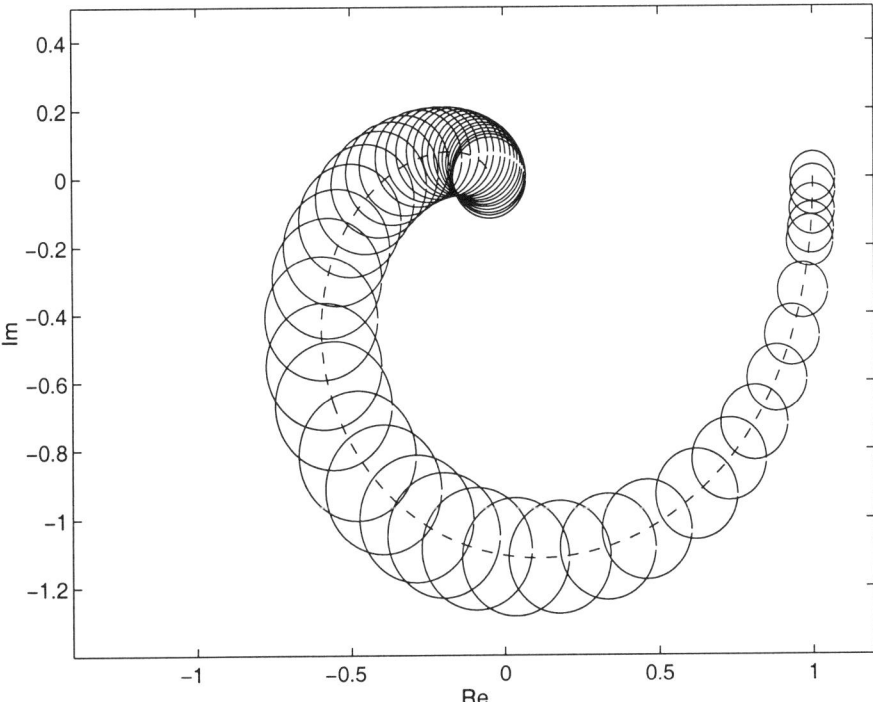

Fig. 15.10. Predicted 1σ bound of the Nyquist plot (second-order plant and second-order noise).

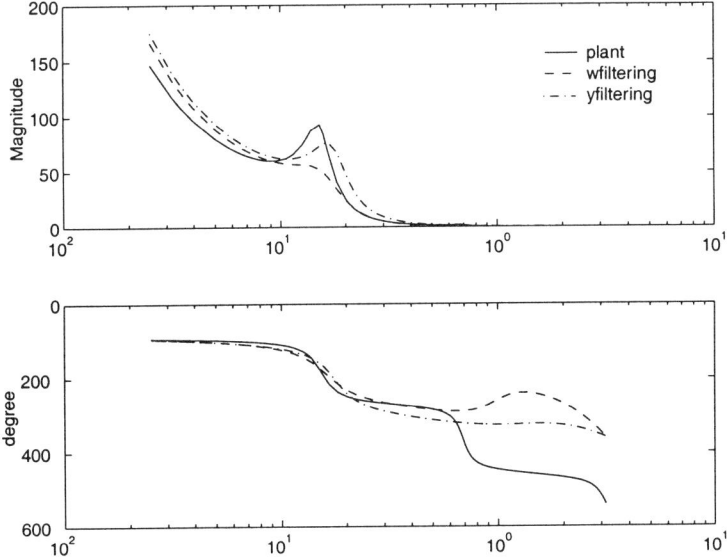

Fig. 15.11. Comparison between y-filtering and w-filtering approaches.

204   15. Closed-loop Identification

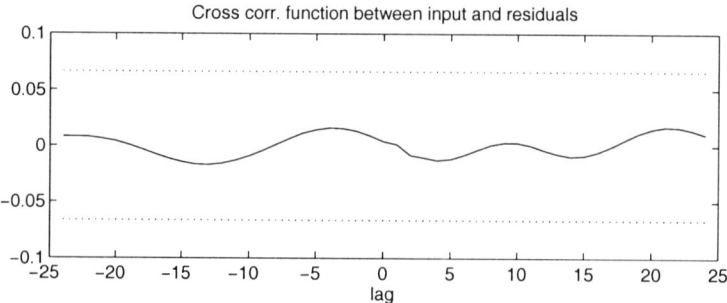

Fig. 15.12. Cross-correlation test for w-filtering.

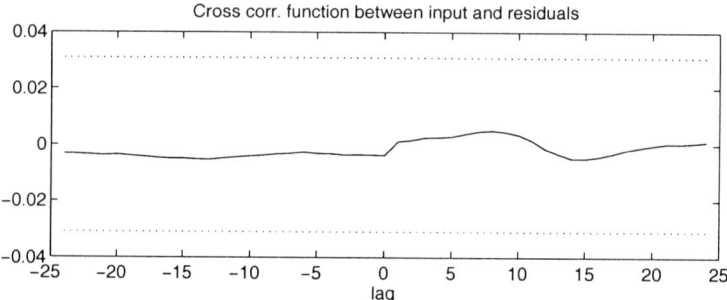

Fig. 15.13. Cross-correlation test for y-filtering.

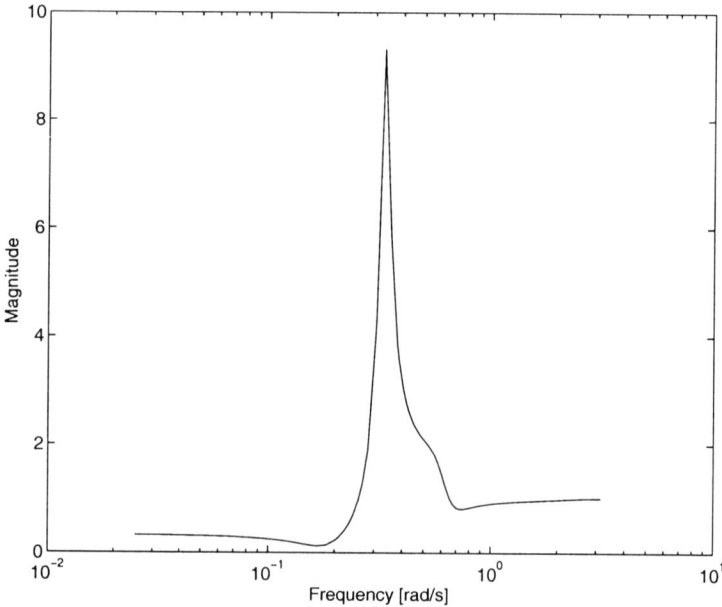

Fig. 15.14. Estimate of the sensitivity function

15.5 Simulation   205

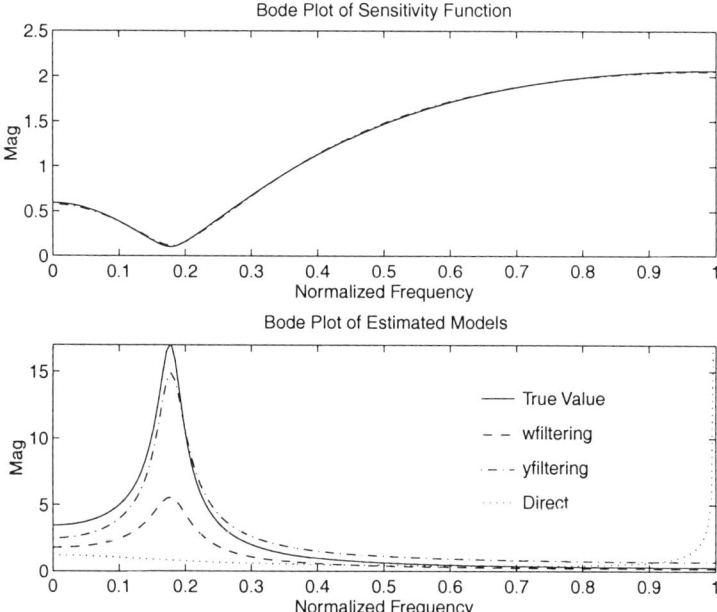

Fig. 15.15. The upper plot is the sensitivity function. The lower plot is the averaged Bode magnitude graph of $\hat{T}$ over 50 runs.

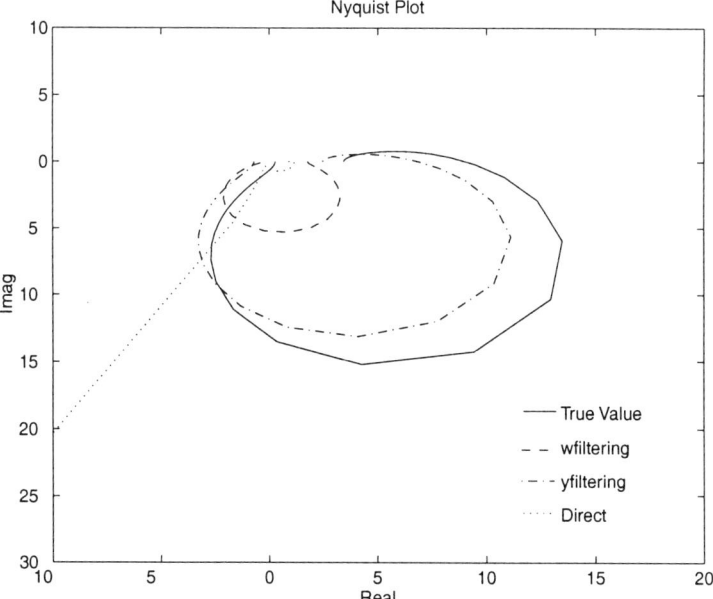

Fig. 15.16. The averaged Nyquist plot of the estimate over 50 runs

15.6 Experimental evaluation on a pilot-scale process

In practical situations, it is difficult to validate the model, $\hat{T}$, estimated under closed-loop conditions with the real process, T, since the latter is unknown. In the following experimental study, separate identification tests are performed under open-loop and closed-loop conditions for the purpose of practical evaluation. The model estimated under closed-loop conditions can be considered suitably adequate and validated if it matches the model estimated under carefully designed open-loop conditions.

Example 15.6.1. The proposed algorithm is evaluated on a pilot-scale process shown in Figure 15.17. Each tank is a double-walled glass tank 50 cm high with an inside diameter of 14.5 cm. The level of the second tank is the output or controlled variable. The water flow to the first tank is manipulated in order to control the level of the second tank. A PID controller (with $T_s = 1$ s) is implemented on the inner loop (flow loop). An IMC controller ($T_s = 5$ s) is implemented on the outer loop (level loop). The block diagram of the real-time SIMULINK Workshop implementation of the IMC controller is shown in Figure 15.18. A second-order model was obtained from an open-loop test. This model was validated by checking it with a separate input-output data set. Closed-loop tests were then conducted. Using the proposed method and other closed-loop identification methods, several process models were obtained. These models were compared with the model obtained from the open-loop test.

Figure 15.19 shows the computer-generated random binary sequence as used in the open-loop test. The step-type random binary sequence was smoothed by a second order Butterworth filter with the cutoff frequency significantly larger than the bandwidth of the process. The bandwidth of the process was estimated from previous open-loop tests. A second-order model was estimated by using the prediction error method and found to be

$$\hat{T} = \frac{0.0023 q^{-2}}{1 - 1.9314 q^{-1} + 0.9323 q^{-2}}$$

where the two-unit time delay is due to a zero order hold and an additional artificially introduced one-unit time delay. The predicted and actual data (using a separate validation or test data with different setpoint excitation inputs) are shown in Figure 15.20. Clearly, the open-loop model is a good representation of the real process.

Since there is integral action in the IMC control, it is preferable to insert an excitation signal via the setpoint to avoid a pole on the unit circle in the sensitivity function decoupling filter. Figure 15.21 shows the excitation signal and the output under the closed-loop test. The y-filtering, w-filtering and direct closed-loop identification methods were used to estimate the process model. Second-order models are identified and the resulting Nyquist plots are

15.6 Experimental evaluation on a pilot-scale process

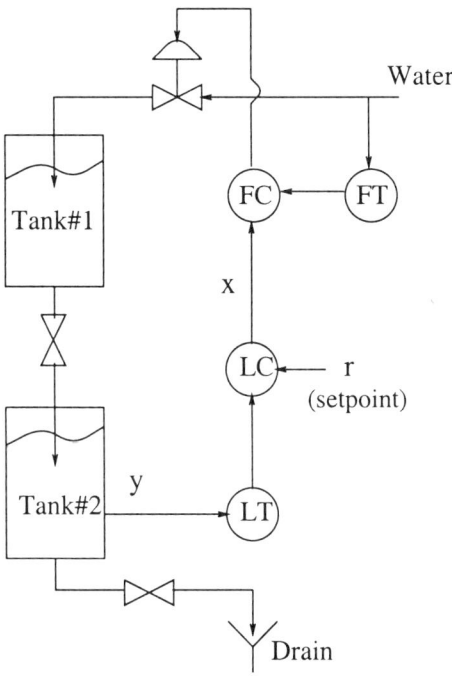

Fig. 15.17. Schematic of the computer-interfaced pilot-scale process.

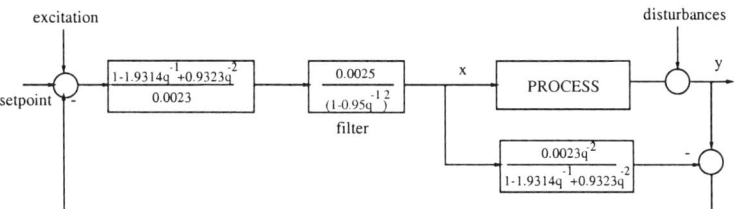

Fig. 15.18. Block diagram for implementation of IMC control using the real-time SIMULINK Workshop.

208   15. Closed-loop Identification

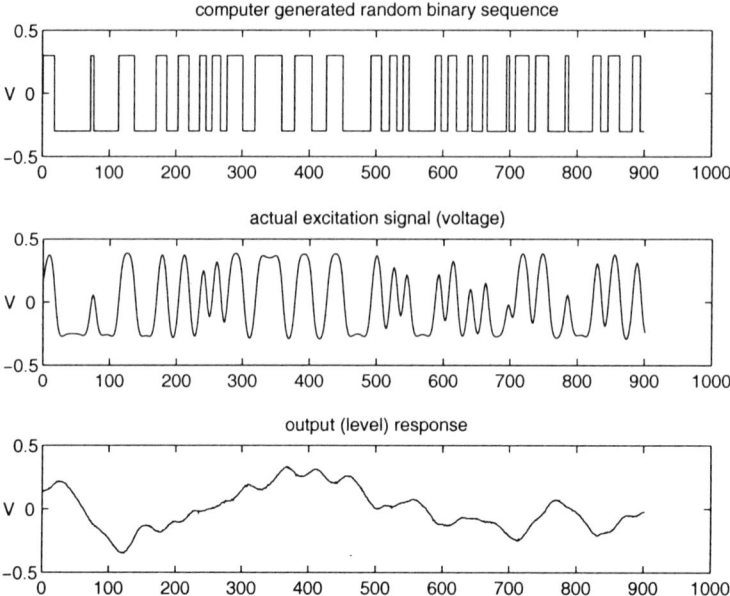

Fig. 15.19. Excitation signal and response. The physical units are voltage in the plot where $-2V$ to $+2V$ correspond to 0% to 100%. The time scale is in terms of sampling intervals.

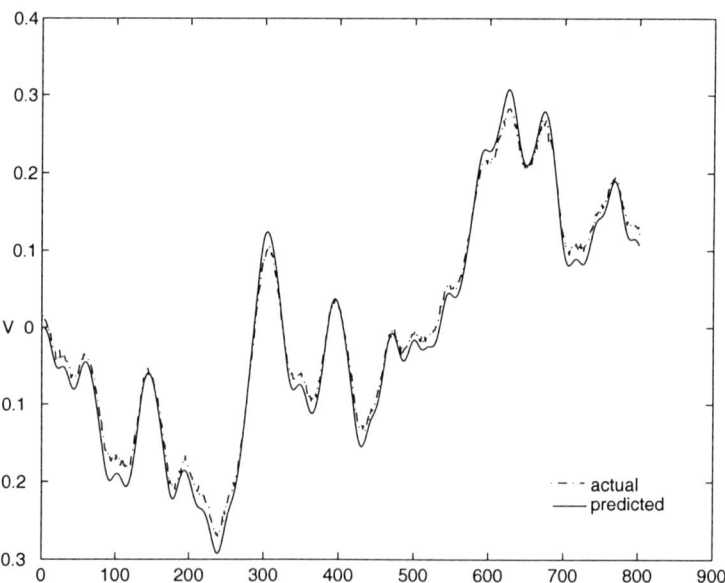

Fig. 15.20. Predicted and actual data from another open-loop test. The time scale is the sampling intervals.

shown in Figure 15.22. If there is no model-plant mismatch, all the Nyquist plots should converge into one plot when the sample size increases.

If a first order process model is assumed, then a model-plant mismatch is indeed present. A larger bias error would be expected at lower frequencies if w-filtering or direct closed-loop identification is used. Nyquist plots of the identified models shown in Figure 15.23 confirm this. Since this is an overdamped second order plant, the model-plant mismatch, which occurred by using a first order model to represent a second-order over-damped plant, is not severe. The direct identification does not fail in this example. The effect

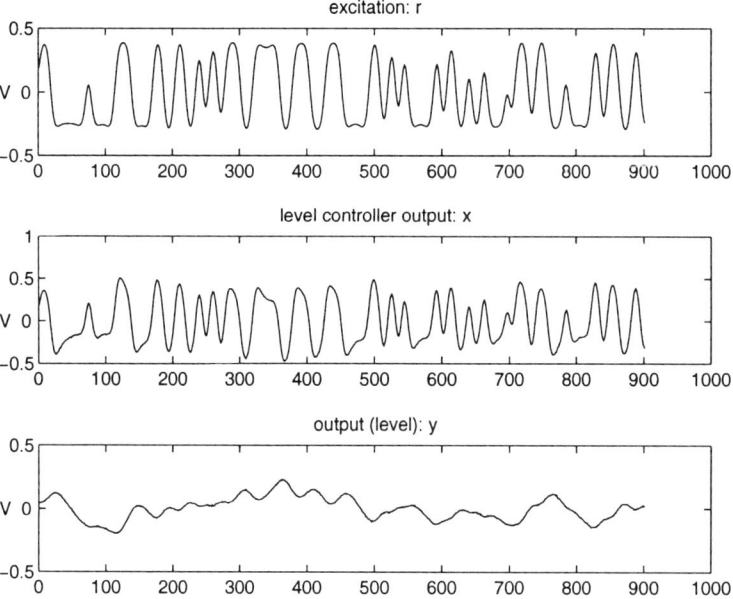

Fig. 15.21. Excitation signal and response under the closed- loop condition. All physical units are voltage in the plot where $-2V$ to $+2V$ correspond to 0% to 100%. The time scale is in terms of sampling intervals.

of the shaping filter on the identification is illustrated by using a fourth-order low-pass Butterworth filter cascaded to the decoupling filter as shown by the results displayed in Figure 15.24. Clear improvement of the estimate at low to middle frequencies is obtained by cascading the shaping filter to the decoupling filter. Depending on the application, different shaping filters at different frequencies can be designed for the control algorithm of choice.

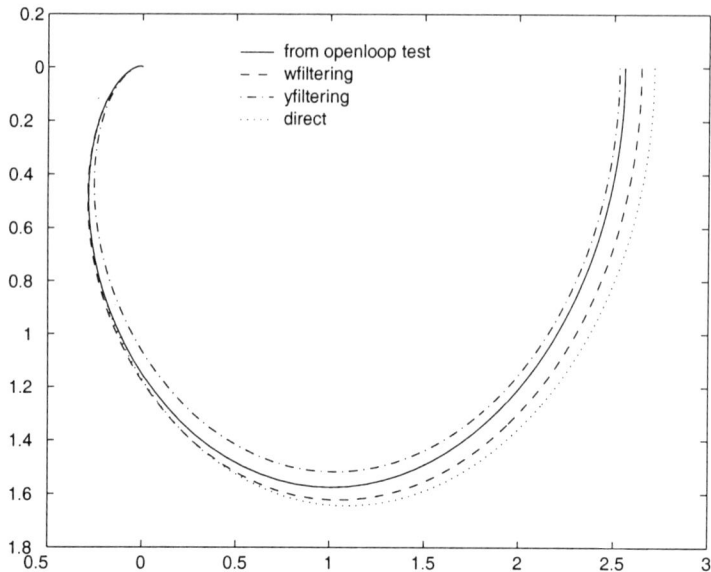

Fig. 15.22. Comparison of the identified process models using different methods when a second-order model is used.

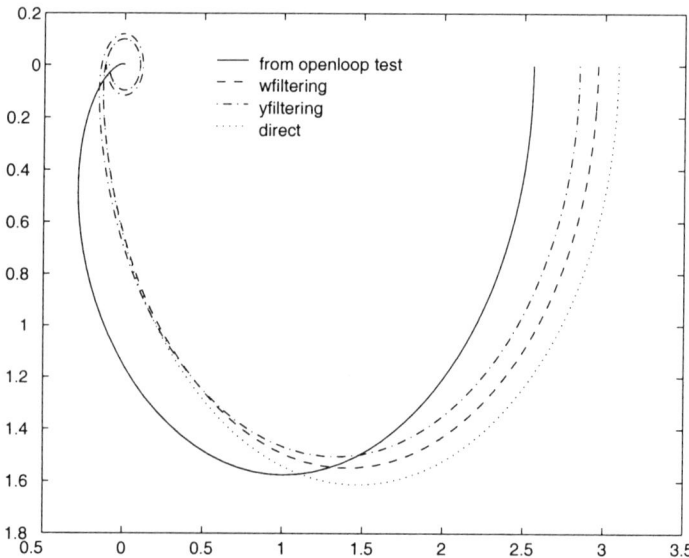

Fig. 15.23. Comparison of the identified process models using different methods when a first-order model is used.

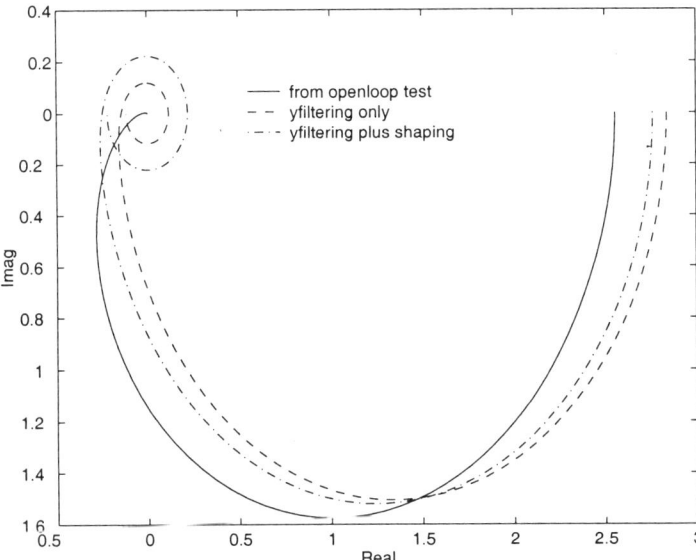

Fig. 15.24. Effect of the shaping filter for the first-order model.

15.7 Summary

The accuracy aspects of closed-loop identification have been discussed. It has been shown that the key difference between closed-loop and open-loop identification is the sensitivity function. The sensitivity function affects the variance and bias errors of the estimate under closed-loop conditions inversely. A two-step closed-loop identification has been proposed, which yields asymptotic properties identical to those under open-loop identification. The proposed algorithm has been evaluated by simulated examples as well as by pilot-scale experiments. These results affirm the strategy that a suitable model commensurate with its intended end use can always be identified under closed-loop conditions through the choice of appropriate data prefilters.

CHAPTER 16
PRATICAL CONSIDERATIONS AND INDUSTRIAL CASE STUDIES

16.1 Introduction

It has now been well recognized that continuous performance assessment is essential for maintaining the advanced process control (APC) assets in the process industry. The commissioning of elaborate control system platforms, e.g., DCS (distributed control system), advanced control applications, e.g., Model-based Predictive Control (MPC), and Information management systems, e.g., Historians and Databases *etc.*, have become commonplace in the process industry. Incidentally, these investments have led to the accumulation of tremendous amounts of process data with few data-mining tools and control-relevant techniques for extracting information.

The implementation of an advanced control strategy is not an end in itself. Continuous improvement in process performance must be ensured by constantly assessing the performance of the basic control loops. Recent studies show that significant improvements in performance, up to 30% reduction in variance, can be realized by re-tuning most of the basic control loops in the process industry (Bialkowski, 1993). Recognizing the importance of loop maintenance, several manufacturing companies have recently undertaken control loop performance assessment for control loops in several plants. Researchers and process control engineers are able to use routine closed-loop plant data to calculate control loop performance indices and perform loop diagnosis. These analyses have led to significant improvement in control loop performance in these plants, and the results are reported in this chapter.

The distinct feature of performance assessment, using minimum variance control as a benchmark, is that the performance index can be calculated by using only routine operating data with *a priori* knowledge of time-delays or interactor matrices. Although this technique has proven to be very useful in many applications, expertise and experience are required for appropriate data preprocessing, performance calculation, and interpretation of results. This requirement limits its application in the plant environment. In this chapter, step by step illustrative examples and explanations of performance assessment and diagnosis are discussed. We hope that this will help readers to apply the theory discussed in the previous chapters in practice.

214   16. Practical Considerations and Industrial Case Studies

16.2 Practical considerations in univariate performance assessment

As has been discussed in the previous chapters, an estimate of the performance index is obtained by whitening the process output via time series analysis and subsequent correlation analysis. In practice, however, *a priori* knowledge of time delays may not be always available. It is therefore useful to assume a range of time delays and then calculate performance indices over this range of the time delays. The indices over a range of time delays are also known as extended horizon performance indices (Thornhill, Oettinger, and Fedenczuk 1998) and defined here as a performance indices curve. Through pattern recognition, it may be easier to assess performance of the control loops by visualizing the patterns of the performance indices versus time delays. There is a clear relationship between the performance indices curve and the impulse response curve of the control loop.

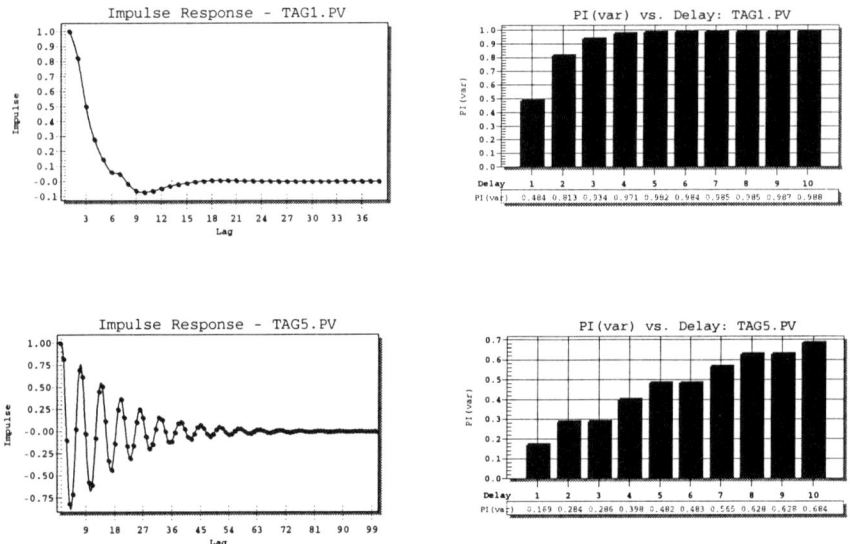

Fig. 16.1. Impulse response subject to impulse disturbance

Let us first consider a simple case where the process is not subject to random disturbances. Figure 16.1 is one example of performance evaluation for a control loop subject to (deterministic) impulse disturbance. This figure shows closed-loop impulse responses (left) and the corresponding performance indices curve (right) calculated from the SSE of the impulse responses. Impulse responses are simulated from a control loop with two different controller tunings. By visualizing the impulse responses, one can find that the

16.2 Practical considerations in univariate performance assessment

closed-loop response of the first loop tuning (denoted as TAG1.PV) has better performance; the closed-loop response of the second loop tuning (denoted as TAG5.PV) has an oscillatory behavior, indicating a relatively poor control performance. With performance index '1' indicating the best possible performance and index '0' indicating the worst performance, performance indices of the first loop tuning (shown on the upper-right plot) approach '1' within 4 time lags, while performance indices of the second loop tuning (shown on the bottom-right plot) take 10 time lags to approach '0.7'. In addition, performance indices of the second loop tuning show ripples as they approach an asymptotic limit, indicating a possible oscillation in the loop. In practice, such clear impulse response curves as those shown in Figure 16.1 are not available since processes are subject to numerous random disturbances. For example, closed loop responses of the same process subject to random white-noise disturbances for the same two set of loop tunings are shown in Figure 16.2. One cannot rank performance of these two controller settings from these noisy data easily. Instead, one may calculate performance indices over a range of time delays (from 1 to 10). The result is shown on the right column plots of Figure 16.2, and the same trajectories of the performance indices as in the previous case are observed. It is evident from these plots that the performance indices curve depends on the controller tuning and dynamics of the disturbances irrespective of the deterministic or stochastic nature of the disturbances.

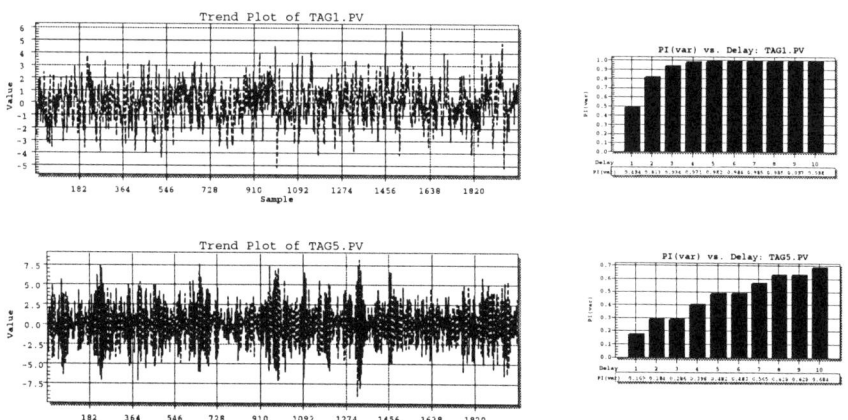

Fig. 16.2. Typical process data and corresponding performance indices

16.3 Univariate performance assessment using alternative performance indicators

In addition to the performance index as a measure for performance assessment, there are several alternative indicators of control loop performance. These are discussed next.

Autocorrelation Function: The autocorrelation function (ACF) of the output, shown in Figure 16.3, is an approximate measure of how close the existing controller is to minimum variance condition, or how predictable the data is over the time horizon of interest. If the controller is in minimum variance condition, then the autocorrelation function should decay to zero after 'd-1' lags where 'd' is the delay of the process. In other words, there should be no predictable information beyond time lag d-1. The rate at which the autocorrelation decays to zero after 'd-1' lags indicates how close the existing controller is to the minimum variance condition. Since it is relatively simple to calculate autocorrelation using process data, the autocorrelation function is often used as a first-pass test before carrying out further performance analysis.

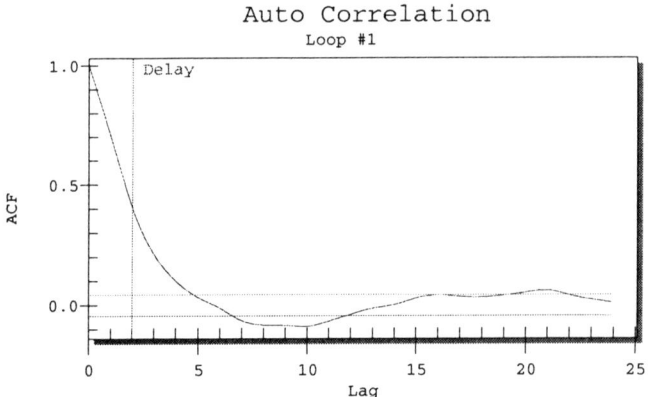

Fig. 16.3. Autocorrelation function

Impulse Response: An impulse response function curve represents a dynamic relationship between the whitened disturbance sequence and the process output. This function is a direct measure of how well the controller is performing in rejecting disturbances or tracking setpoint changes. Under a stochastic framework, this impulse response function may be calculated using time series analysis. Once a time series model (typically an ARMA model) is estimated, the infinite-order, moving average representation of the model can be obtained through a long division of the time series model. Figure 16.4 shows closed-loop impulse responses of a control loop with two different control tunings. Clearly they denote two different closed-loop dynamic responses: one is slow and smooth, and the other one is relatively fast and slightly oscillatory. The sum of square of the impulse response coefficients is the variance of the data.

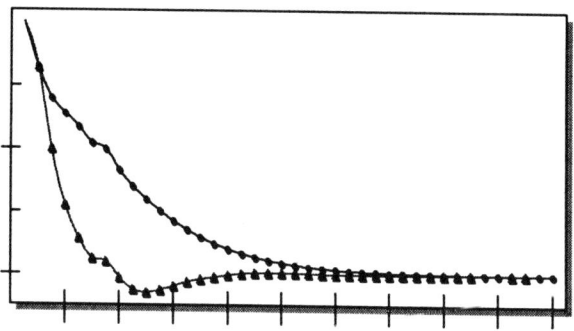

Fig. 16.4. Impulse responses estimated from routine operating data

Spectral Analysis: The closed-loop frequency response of the process data is an alternative way to assess control loop performance. Spectral analysis of output data allows one to detect oscillations, offsets and measurement noises present in the process easily. The closed-loop frequency response is often plotted together with the closed-loop frequency response under minimum variance control. This is to check the possibility of performance improvement through controller tunings. The comparison gives a measure of how close the existing controller is to the minimum variance condition. In addition, it also provides the frequency range in which the controller deviates significantly from minimum variance condition. Typically, a large deviation in the low frequency range indicates lack of integral action or weak proportional gain. Typically, large peaks in the middle frequency range indicate an over-tuned controller or the presence of oscillatory disturbances. As an illustrative example, frequency responses of two control loops are shown in Figure 16.5. The top graph of the figure shows that closed-loop frequency response of the existing controller is almost the same as the frequency response under minimum variance control. A peak at the mid-frequency may indicate a slightly over-tuned controller. The bottom graph of Figure 16.5 shows that the frequency response of the existing controller is oscillatory, possibly indicating a poorly-tuned controller or the presence of an oscillatory disturbance; otherwise the controller is close to minimum variance condition.

Residual test: The algorithm used for performance analysis is based on the assumption that the system is linear or approximately linear. Residuals

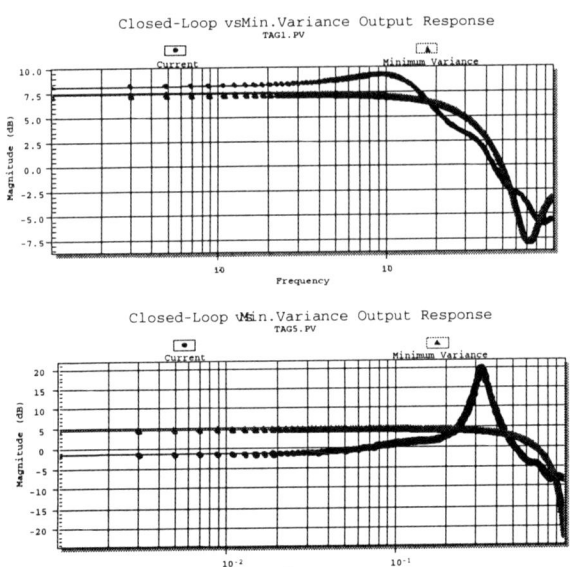

Fig. 16.5. Frequency response estimated from routine operating data

are obtained by fitting a linear model to the process data. If the residuals pass the whiteness test, then the estimated models that are used for performance assessment are valid. Figure 16.6 shows an illustrative example where the residuals pass the whiteness test. If the autocorrelation function of the residuals lies between 95% confidence intervals, then the residuals are regarded as white and pass the test. Residuals that pass the whiteness test are also called innovation sequences. Innovation sequences play an important role in the FCOR algorithm as discussed in the previous chapters.

Segmentation of performance indices: Many process data exhibit time varying dynamics, *i.e.*, the process transfer function or the disturbance transfer function is time variant; therefore, performance assessment with a non-overlapping sliding data window that can track time-varying dynamics is often desirable. For example, segmentation of data may lead to some insight into any cyclical behavior of the process variation in controller performance during, for example, day/night or due to shift change *etc.*, Figure 16.7 is an example of performance segmentation over a window of 200 data points.

16.3 Univariate performance assessment: alternative indicators 219

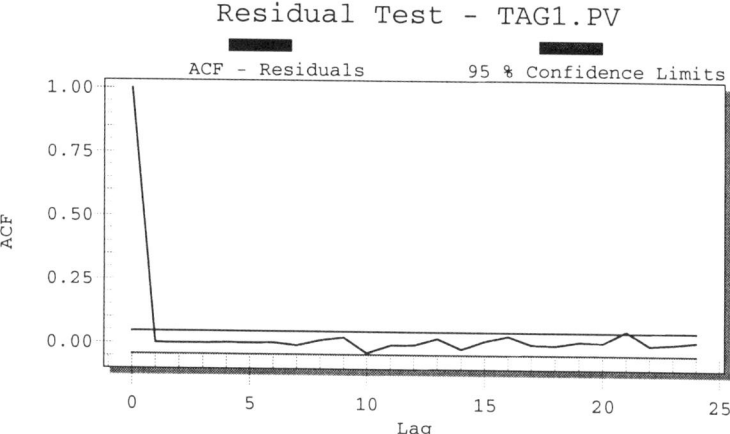

Fig. 16.6. Autocorrelation of residuals

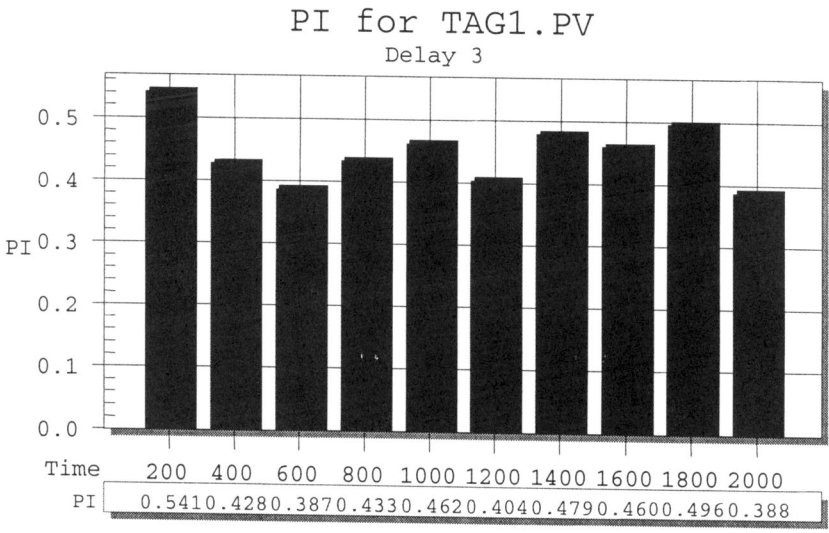

Fig. 16.7. Segmentation of performance indices

16.4 Univariate performance assessment using user specified benchmarks

The increasing level of global competitiveness has pushed chemical plants into high-performance operating regions that require advanced process control technology. Consequently, industry has an increasing need to upgrade its conventional PID controllers to advanced control systems. The most natural questions to ask for such an upgrading are follows. Has the advanced controller improved the performance as expected? If yes, where is the improvement and can it be justified? Has the advanced controller been tuned to its full capacity? Can this improvement also be achieved by simply re-tuning the existing traditional, e.g., PID, controllers? In other words, what is the cost versus benefit of implementing an advanced controller? Unlike performance assessment using minimum variance control as benchmark, the solution to this problem does not require *a priori* knowledge of time-delays. Two possible benchmarks may be chosen: one is the historical data benchmark or reference data set benchmark, and the other is a user-specified benchmark. The purpose of reference data set benchmarking is to compare performance of the existing controller with the previous controller during the 'normal' operation of the process. This reference data set may represent the process when the controller performance is considered satisfactory with respect to meeting the performance objectives. The reference data set should be representative of the normal conditions at which the process is expected to operate at, *i.e.*, the disturbances and set-point changes entering into the process should not be unusually different. This analysis provides the user with a relative performance index (RPI) which compares the existing control loop performance with a reference control loop benchmark chosen by the user. The RPI is bounded by $0 < RPI < \infty$, with '< 1' indicating deteriorated performance, '1' indicating no change of performance, and '> 1' indicating improved performance. Figure 16.8 shows a result of reference data set benchmarking. The impulse response of the reference data (denoted as 'Benchmark' in the figure) decays to zero smoothly, indicating good performance of the controller. After one increases the proportional gain of the controller, the new impulse response (denoted as 'Current' in the figure) shows oscillatory behavior, with an RPI=0.4, indicating deteriorated performance owing to the oscillation.

In some cases, one may wish to specify certain desired closed loop dynamics and carry out performance analysis with respect to such desired dynamics. One such desired dynamic benchmark is the closed loop settling time discussed in the previous chapters. As an illustrative example, Figure 16.9 shows a system where a settling time of 10 sampling units is desired for a process with a delay of 5 sampling units. The impulse responses show that the existing loop is close to the desired performance, and the value of RPI=0.9918 confirms this so no further tuning of the loop is necessary.

16.4 Univariate performance assessment using user specified benchmarks   221

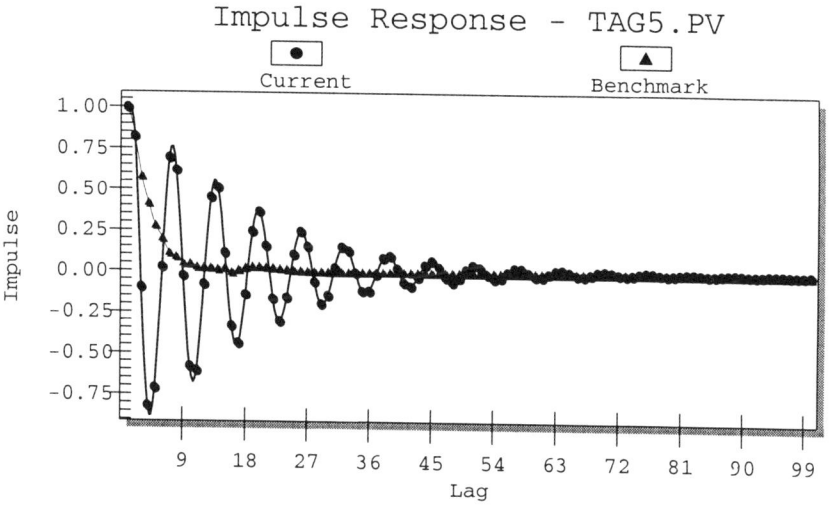

Fig. 16.8. Reference benchmarking based on impulse responses

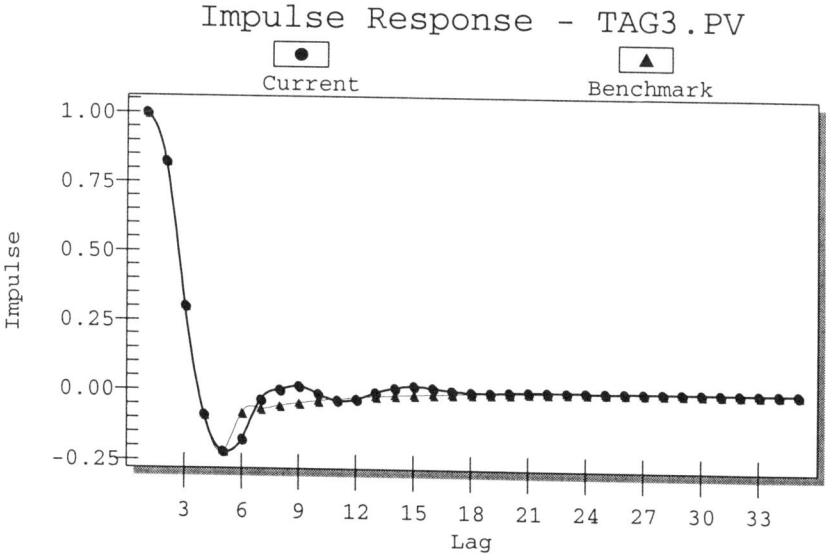

Fig. 16.9. User specified benchmark based on impulse responses

16.5 Performance assessment of multivariate control loops

16.5.1 Performance assessment of multivariate control loops using minimum variance control as benchmark

Performance assessment of univariate control loops is carried out by comparing the actual output variance with the minimum output variance. The latter term is estimated by simple time series analysis of routine closed-loop operating data. This idea has been extended to multivariate control loop performance assessment and the multivariate FCOR algorithm has been developed in the previous chapters.

A multivariate performance index is a single scalar measure of multivariate control loop performance relative to its corresponding output under multivariate minimum variance control. Individual output performance indices indicate performance among each output relative to the multivariate minimum variance control. If a particular output index is much smaller than other output indices, then some of the other loops may have to be de-tuned in order to improve this poorly tuned loop. Figure 16.10 is an example of performance assessment of the Wood-Barry distillation column control system. It can be observed from this figure that the multivariate performance index=0.75. The index of the first output=0.6897 and the index of the second output=0.9356. Since this is a multivariate control system under multiloop PID control, the relatively low index of the first output indicates a relatively poorly-tuned controller, while the high index of the second output indicates a well-tuned controller for this loop. Since the multivariate performance index must be no more than one, an increase in one output's individual index may lead to a decrease in another's. Note that the individual output index is not necessarily bounded by one.

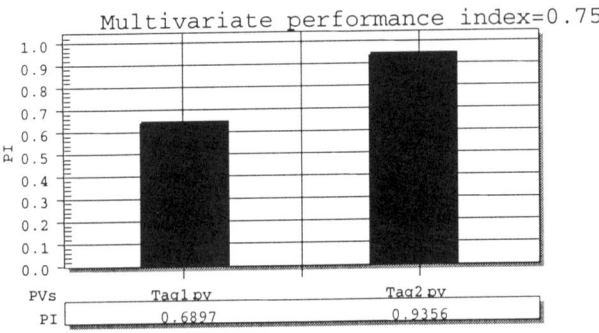

Fig. 16.10. Multivariate performance assessment of a simulated distillation column control system

16.5.2 Performance assessment and diagnosis of multivariate control loops using alternative performance indicators

Autocorrelation Function: The autocorrelation function (ACF) plots can be used to analyze individual process output performance. A typical example of the ACF plots for the two output variables of the simulated Wood-Barry column control system is shown in Figure 16.11. The diagonal plots are autocorrelations of each output variable, while the off-diagonal plot is cross-correlation. Typically, the diagonal plots indicate how well each loop is tuned. For example, a slowly decaying autocorrelation function implies an under-tuned loop, and an oscillatory ACF implies an over-tuned loop. Off-diagonal plots can be used to detect the interaction between each process outputs. Figure 16.11 clearly indicates that the first loop has relatively poor performance while the second loop has very fast decay dynamics and thus, good performance. Interaction between the two loops can also be observed from the off-diagonal plot.

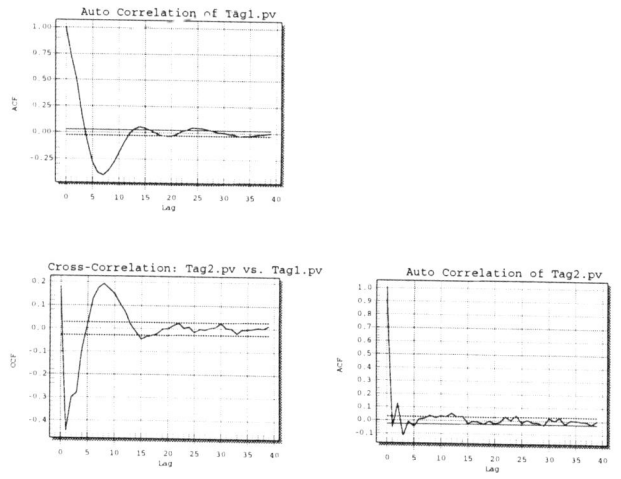

Fig. 16.11. Autocorrelation function of multivariate process

Frequency Response: Frequency domain plots or spectrum plots provide alternative indicators of control loop performance. They may be used to assess individual output dynamic behavior, interactions and effects of disturbances. Typically, for example, peaks in the diagonal plots imply oscillation of the output variables owing to an over-tuned controller or presence of oscillatory disturbances. Frequency domain plots also provide information on the frequency ranges over which the oscillations occur and the amplitude of the oscillations. Like time domain analysis, off-diagonal plots provide one with information on the correlation or interaction between the loops. Figure 16.12 is the frequency response plot of the simulated Wood-Barry column. The first diagonal plot indicates that there is a clear mid-frequency oscillation in the 1st output and some oscillation in the 2nd output as evident from the second diagonal plot. Off-diagonal plot shows there is an interaction at the above-mentioned oscillation frequency.

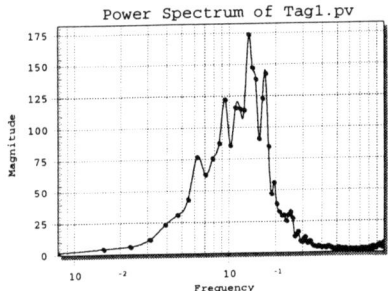

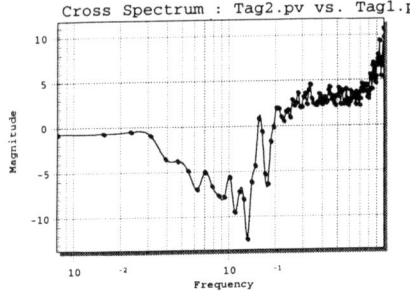

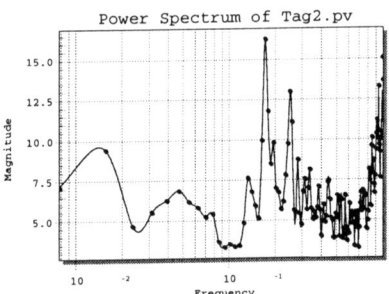

Fig. 16.12. Frequency response of multivariate process

Normalized Multivariate Impulse Response: An impulse response curve represents a dynamic relationship between the whitened disturbance and the process output. Alternatively, one may treat this curve as what constitutes the process response if the 'shock' of disturbance is a single impulse. In the univariate case, the first 'd' impulse response coefficients are feedback controller invariant, where 'd' is the process time-delay; therefore, if the loop is under minimum variance control, the impulse response coefficients should be zero after 'd-1' lags. The same idea can be applied to a multivariate system, except that data has to be transferred to a different co-ordinate in order to determine the feedback controller invariant impulse response coefficients. The Normalized Multivariate Impulse Response (NMIR) curve reflects this idea. The NMIR curve is calculated from Equation 8.5 that is re-written here as

$$\tilde{Y}_t - E(\tilde{Y}_t) = \underbrace{F_0 a_t + F_1 a_{t-1} + \cdots + F_{d-1} a_{t-d+1}}_{e_t}$$
$$+ \underbrace{L_0 a_{t-d} + L_1 a_{t-d-1} + \cdots}_{w_{t-d}} \qquad (16.1)$$

Then the first NMIR coefficient is given by $tr(F_0 \Sigma_a F_0^T)$, the second NMIR coefficient is given by $tr(F_1 \Sigma_a F_1^T)$, and so on.

The first 'd' NMIR coefficients are feedback controller invariant, where 'd' is the maximum time-delay among the elements of the interactor matrix. If the loop is under multivariate minimum variance control, then the NMIR coefficients should decay to zero after 'd-1' lags. The sum of the squares under the NMIR curve is equivalent to the trace of the covariance matrix of the data. The multivariate performance index is equal to the ratio of the sum of the squares of the first 'd' NMIR coefficients and the sum of the squares of all NMIR coefficients. For the simulation example of the Wood-Berry column, the normalized multivariate impulse response is calculated and plotted in Figure 16.13. Since $d = 15$, the first 15 NMIR coefficients are feedback control invariant and depend solely on the disturbance dynamics and the interactor matrix. The sum of the squares of these 15 coefficients is the variance achieved under multivariate minimum variance control. In this simulation example, one can observe that the NMIR decays to zero quickly after the 15 sample units, indicating relatively good overall multivariable control performance.

16.6 Industrial case studies of continuous performance assessment

16.6.1 Composition control loop performance assessment

The main objective in conducting this case study is to demonstrate the value-added benefits that will accrue from Continuous Performance Assessment (CPA). The control loop of interest is a composition controller that maintains the composition of SO_2/H_2S from the SRU at a stipulated setpoint value. The performance index calculations are based on historical routine operating data, with the following implicit assumptions: approximately 10 hours of data sampled at 1 minute time intervals; no setpoint changes in this time interval, hence assessment is based on controller reaction to disturbance; no offset

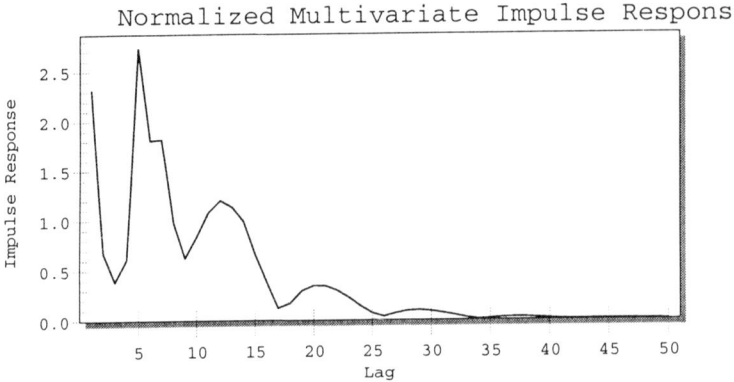

Fig. 16.13. Normalized multivariate impulse response

is found in the data, *i.e.*, mean[SP - PV]=0. It is expected that variance reduction will improve performance. The composition controller is a digital PID controller

The implementation of performance assessment on the composition controller loop at this site is conducted in two phases:

1. **Advanced Data Analysis:** collect process data from the Historian, preprocess and analyze process data; perform loop assessment calculations and generate performance audit report.
2. **Implementation of Recommendations:** tune PID loop as suggested in the audit report, preprocess and analyze newly-acquired process data; perform relative performance assessment calculations and generate performance audit report.

Figure 16.14 shows the auto-correlation of the process output error. The output error is defined as the difference between the setpoint (SP) and the process variable (PV) or [SP - PV]. If the controller is close to the minimum variance condition then the autocorrelation function should decay to zero after d-1 lags where d = 'delay'. From Figure 16.14 it can be seen that the autocorrelation function decays very slowly, indicating that the controller is not close to the minimum variance condition.

The performance index for the composition controller (assuming a delay of 3 minutes) was 0.45, but as is often the case with industrial settings, the delay is not accurately known so the performance index is obtained using a range of delays (1 - 5). The plot of process delay versus computed performance index values is shown in Figure 16.15.

The performance index may also vary with changes in the nature of disturbances in the process. In order to determine whether there is a changing disturbance structure over the time span of the data, the performance index

16.6 Industrial case studies of continuous performance assessment 227

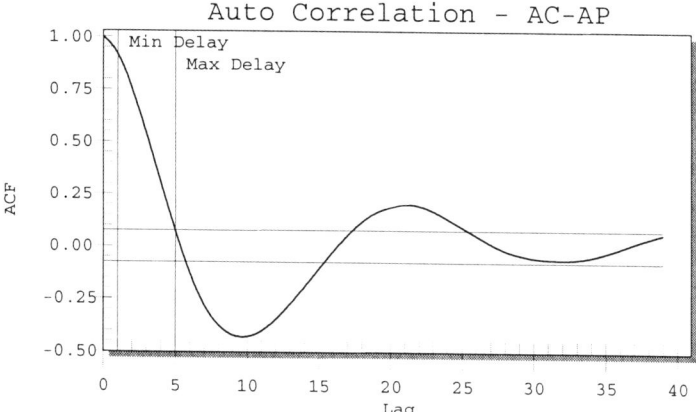

Fig. 16.14. Auto-correlation function of the composition controller

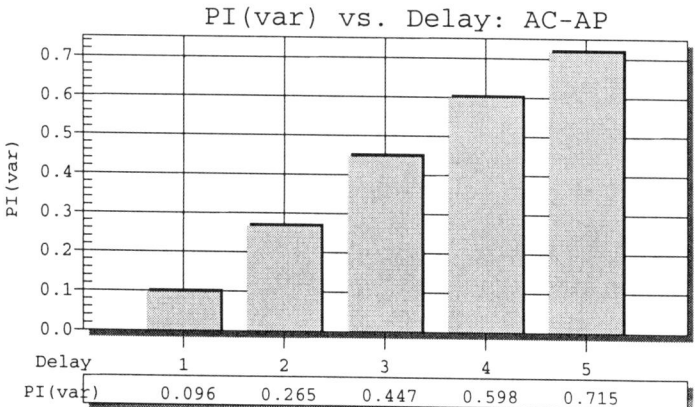

Fig. 16.15. Performance index for a range of delays (1 - 5 minutes)

is obtained for non-overlapping 3-hour data segments (180 samples). This is shown in Figure 16.16.

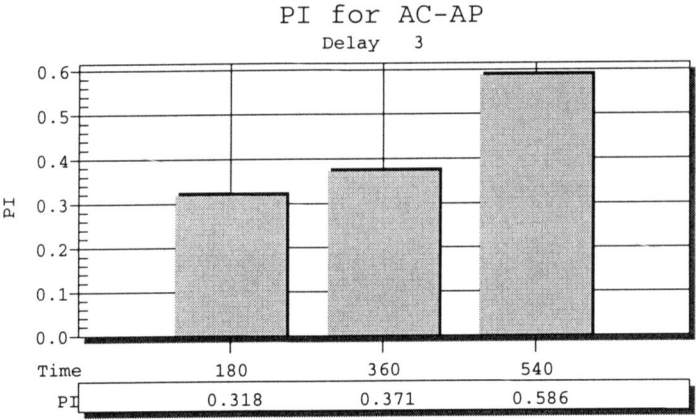

Fig. 16.16. Performance index for a moving window of size 180 samples

It is observed from Figure 16.15 that the performance index increases slowly with the time delay. This performance index has been considered as being relatively low for a critical composition loop. Figure 16.16 shows the time-varying performance index, but overall the index is low. Furthermore, the closed-loop impulse response shown in Figure 16.17 indicates the presence of mild under-damped oscillations with quarter decay ratio. The frequency domain response plots in Figure 16.18, show the presence of a spike (peak) at a frequency of 0.1. This frequency axis has been scaled relative to the Nyquist frequency (1/2 sampling frequency).

The following recommendations for this loop are obtained from software that performs this analysis:

– Output variance may be reduced by re-tuning the existing controller.
– Reduce the integral gain (or equivalently, increase the integral time constant/reset time).

The preceding analysis of historical control loop data indicates that performance improvement may be achieved by retuning the controller. This recommendation is implemented in the DCS, by increasing the integral time constant from 0.40 minutes/cycle to 0.45 minutes/cycle. Segments of the data collected for both PV and SP over the next 24 hours are then used to evaluate the RPI for the newly retuned loop. An RPI value of 1.2404 is obtained, with a PI value of 0.6729, which is higher than the previous value of 0.45; hence, implementing the recommendation suggested by the software resulted in significant performance improvement for the composition loop. In

16.6 Industrial case studies of continuous performance assessment   229

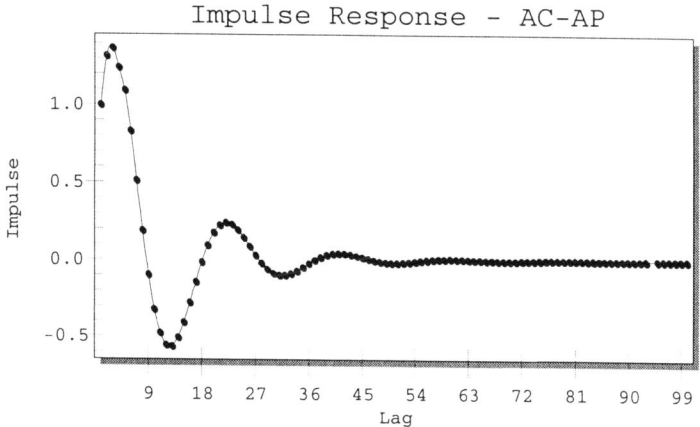

Fig. 16.17. Closed loop impulse response coefficients of the composition controller response (PI=0.45, d=3 min)

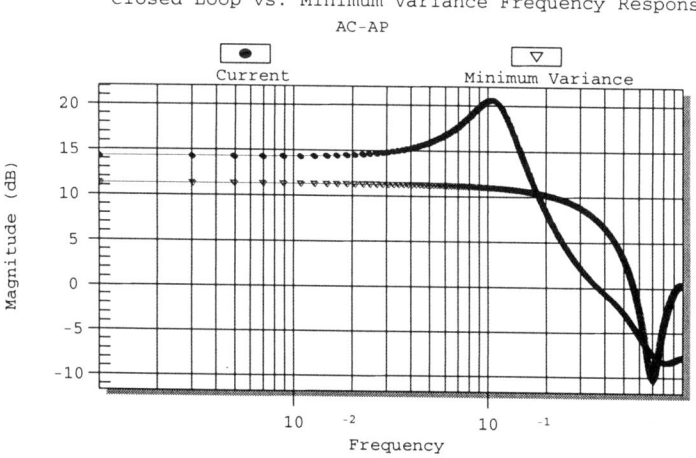

Fig. 16.18. Closed loop frequency response plots of the composition controller response (PI=0.45, d=3 min.)

addition, the impulse response plot in Figure 16.19 also shows significant attenuation of the under-damped, oscillatory response that is observed before retuning the controller. Comparisons of the frequency domain response plots in Figure 16.20 show a significant improvement in the closed-loop response in the mid-frequency range.

In summary, the results of this case study have demonstrated that significant economic benefits can accrue from conducting control loop performance assessment. The extension of this technique to continuous performance assessment can be appreciated from the point of view that if control loops with degrading performance can be identified and retuned quickly, considerable improvement in overall performance and economic benefits can be realized in the plant on a continuing basis.

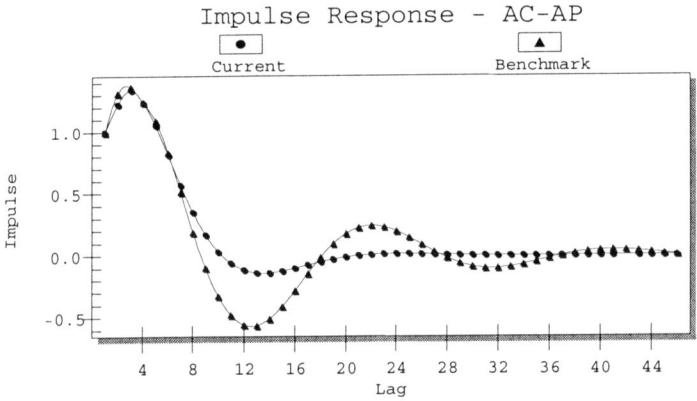

Fig. 16.19. Comparisons of the closed impulse response plots of the composition controller response before and after tuning

16.6.2 Capacitance drum control loops performance assessment

Capacitance drum control loops of a plant (denoted here as Plant AB) are to be studied. The primary objective of Plant AB is to reduce the water in the previous plant (Plant A) diluted bitumen product further, prior to it reaching the next plant (Plant B) storage tanks. As the grade of the feed entering plant A reduces, the water required to process the oil sands increases proportionally. A large portion of this excess water ends up in the Plant A froth feed tank and ultimately, increases the % volume of water in the Plant A product. Aside from degrading the quality of the product, the increased volume of water means reduction in the amount of bitumen that can be piped to the diluted bitumen tanks. In addition, the higher water content means more of the chloride compound present in the oil sands is dissolved

16.6 Industrial case studies of continuous performance assessment

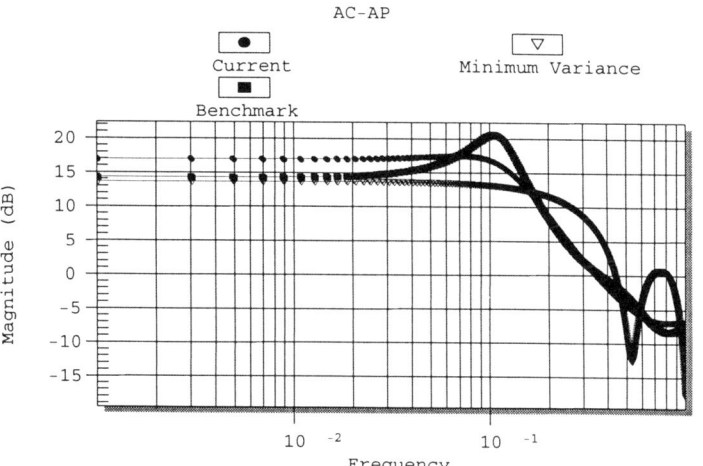

Fig. 16.20. Comparisons of the closed loop frequency response plots of the composition controller response before and after tuning

and finds its way to the diluted bitumen tanks and eventually to plant B. The higher chloride concentration increases the corrosion rate of equipment in the Upgrading units. Plant AB was developed as a means of reducing the water content, and ultimately, the chlorides sent to Upgrading. This reduction is achieved by centrifuging the Plant A Product. All product from plant A is directed to the Plant AB feed storage tank. The IPS portion of the product is routed through 5 Cuno Filters prior to entering the Plant AB feed storage tank. Feed from the feed storage tank is then pumped through the feed pumps to the Alfa Laval centrifuges. The Alfa Laval centrifuges remove water and a small amount of solids from the feed. Each centrifuge has its own capacitance drum and product back pressure valve. This arrangement allows for individual centrifuge "E-Line" control and a greatly improved product quality. Heavy phase water from plant A is used as Process Water in plant AB.

The cap drum pressure controller (671PC3H) controls the capacitance drum pressure by adjusting the nitrogen flow into the drum. The Cap Drum Primary level controller (671LC1AH) maintains the cap drum water level by adjusting water addition into the drum. Control of these two variables is essential to maintain the E-Line in the centrifuges. Currently these two loops are controlled by multiloop PID controllers. The two process variables, pressure and level, are highly-interacting. The objective of the performance assessment is to evaluate the existing multiloop PID controllers' performance, and to identify opportunities, if any, to improve performance by implementing a multivariate controller

Process data with a five-second sampling interval are shown in Figure 16.21. These are typical (representative) process data encountered in this process. By assuming that both pressure and level loops have no time delay except for the delay induced by the zero-order-hold device, a scalar multivariate performance index was calculated as 0.022 and individual output indices are shown in Figure 16.22. Based on these indices, one may conclude that controller performance is poor and may be improved significantly by retuning the existing controllers or redesigning a multivariate controller, but since the exact time delays for these loops are unknown, further analysis of performance in both the time domain and the frequency domain is necessary. For example, the normalized impulse response shown in Figure 16.23 does indicate that the disturbance persists for about 50 samples before it is completely compensated by the controllers. This is equivalent to a settling time of 4 minutes for the overall system. To check which loop causes such a long settling time, one can look at the auto- and cross-correlation plots. The individual loop behavior can be observed from the auto- and cross-correlation plots shown in Figure 16.24. It is observed that the pressure response does not settle down even after 40 samples. This is clearly unacceptable for a pressure loop. In addition, a significant oscillation is observed in the level response. This oscillation is confirmed from the spectrum plot shown in Figure 16.25. Notice that the peak (oscillation) appears in both the pressure and level responses as well as in the cross-spectrum plot. This indicates that both loops have oscillation and they do interact significantly at that frequency. Thus, this analysis indicates that: 1) the existing multiloop controller has relatively poor performance, primarily due to the long settling time and oscillatory behavior or presence of oscillatory disturbances; 2) the two loops are strongly interacting and a multivariate controller may be able to improve performance significantly.

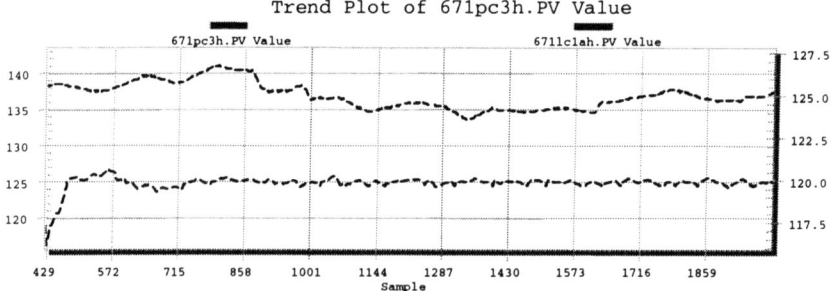

Fig. 16.21. Pressure and level data with sampling interval 5 seconds

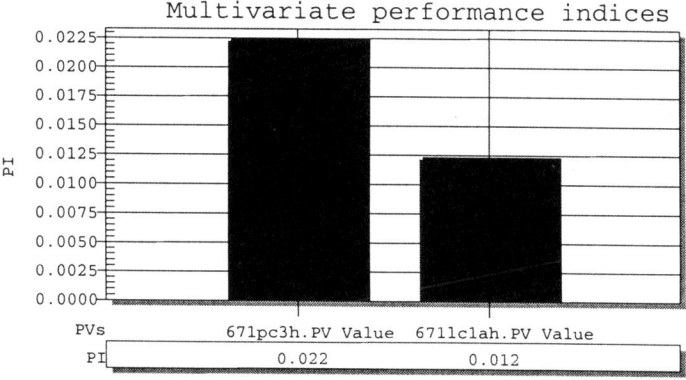

Fig. 16.22. Individual output performance indices

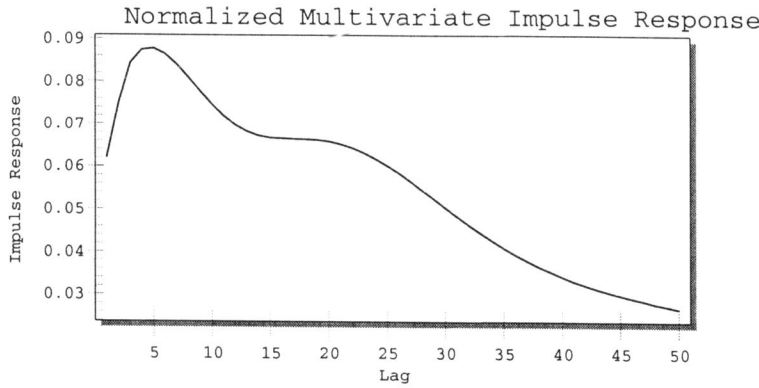

Fig. 16.23. Normalized multivariate impulse response

16.7 Summary

In summary, industrial control systems are designed and implemented or upgraded with a particular objective in mind. The new loop performance assessment methodology proposed here will permit automated and repeated monitoring of the design, tuning and upgrading of the control loops. Poor design, tuning or upgrading of the control loops will be detected, and repeated performance monitoring will indicate which loops should be retuned or which loops have not been effectively upgraded when changes in the disturbances,

234   16. Practical Considerations and Industrial Case Studies

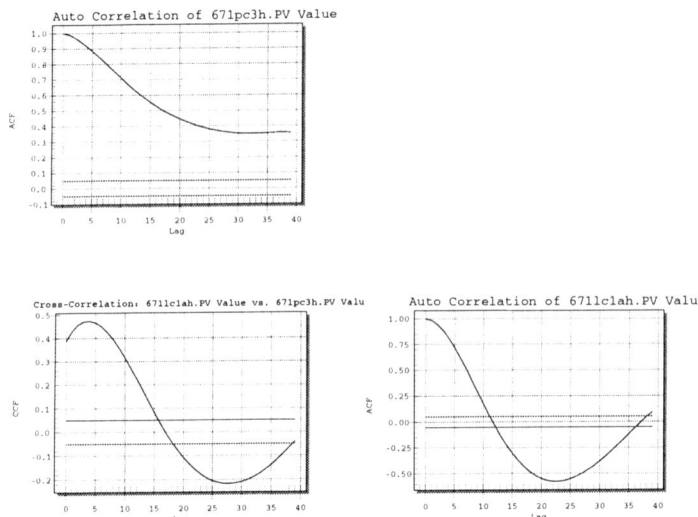

Fig. 16.24. Auto and cross correlation of process output

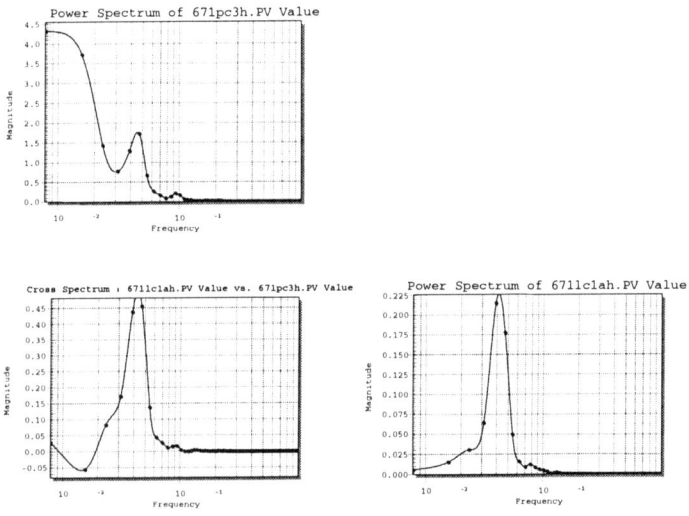

Fig. 16.25. Frequency domain analysis of process output

in the process or in the controller itself occur. Obviously better design, tuning and upgrading will mean that the process will operate at a point close to the economic optimum, leading to energy savings, improved safety, efficient utilization of raw materials, higher product yields, and more consistent product qualities. This chapter has summarized the major steps for control loop performance assessment and industrial implementation experiences in several manufacturing companies. The results have demonstrated applicability of this new technique in industrial process control for enterprise asset management. The study illustrates how controllers, whether in hardware or software form, should be treated like 'capital assets'; how there should be routine monitoring to ensure that they perform close to the economic optimum and that the benefits of good regulatory control are achieved.

APPENDIX A
THE ALGORITHM FOR THE CALCULATION OF A UNITARY INTERACTOR MATRIX

The following algorithm is from Rogozinski *et al.* (1987) and Peng and Kinnaert (1992).

Definition A.0.1. *The $n \times n$ first degree polynomial matrix $U(q)$ will be called a row shift polynomial matrix (r.s.p.m) of order k_i, where*

$$U(q) = U_0 q + U_1 = \begin{bmatrix} 0 & I_r \\ qI_{k_i} & 0 \end{bmatrix}$$

The matrices U_0 and U_1 are defined through the matrix of coefficients

$$U = \begin{bmatrix} U_0 \\ U_1 \end{bmatrix} = \begin{bmatrix} 0_r \\ I_n \\ 0_{k_i} \end{bmatrix}, \quad n = r + k_i$$

in which U_0, U_1 are of dimension $n \times n$, I_n is the $n \times n$ identity matrix, and 0_r is a r-row matrix of zeros.

From RMF (right matrix fraction) description of $T(q^{-1}) = N(q)R^{-1}(q)$, where

$$N(q) = N_0 q^p + N_1 q^{p-1} + \cdots + N_p \qquad (A.1)$$

a block matrix of coefficients is formed as

$$\Lambda = \begin{bmatrix} N_0 \\ \vdots \\ N_p \end{bmatrix}$$

The unitary interactor matrix $D(q)$ can be factored out from Equation A.1 (or the block matrix of coefficients) by the following theorem.

Theorem A.0.1. *(Rogozinski, Paplinski, and Gibbard 1987) For a transfer matrix $T(q)$ satisfying Assumption 1 and 2, there exists a unitary interactor matrix consisting of finite (t) factors:*

$$D(q) = S^{(t)}(q) S^{(t-1)}(q) \cdots S^{(1)}(q) \qquad (A.2)$$

where

$$S^{(i)}(q) = U^{(i)}(q) Q^{(i)} \qquad (A.3)$$

and $U^{(i)}(q)$ is a r.s.p.m. of order k_i and $Q^{(i)}$ is a non-singular $n \times n$ real matrix (an orthogonal matrix for the factorization of the unitary interactor).

A. The algorithm for the calculation of a unitary interactor matrix

The algorithm is as follows:
Set $i = 0$, $N^{(0)}(q) = N(q)$, $\Lambda^{(0)} = \Lambda$, and $D^{(0)} = I_n$ to start the algorithm. Consider the ith iteration in the evaluation of $D(q)$

Step 1:
If $r_i = rank(N_0^{(i-1)}) = min(n, m)$, the algorithm terminates and the unitary interactor matrix is $D(q) = D^{(i-1)}(q)$, set $t = i - 1$;
If $r_i < min(n, m)$, factorize $N_0^{(i-1)}$ by QR factorization into

$$N_0^{(i-1)} = (Q^{(i)})^{-1} \begin{bmatrix} 0_i \\ N_{0D}^{(i)} \end{bmatrix}, \quad i.e., \quad Q^{(i)} N_0^{(i-1)} = \begin{bmatrix} 0_i \\ N_{0D}^{(i)} \end{bmatrix} \quad (A.4)$$

where $Q^{(i)}$ is an $n \times n$ unitary (orthogonal) real matrix, $k_i = n - r_i$ and 0_i is a k_i-row zero matrix.

Step 2:
Pre-multiplying $N^{(i-1)}(q)$ by matrix $Q^{(i)}$

$$\bar{N}(q) = Q^{(i)} N^{(i-1)}(q) \quad (A.5)$$

[the leading coefficient of $\bar{N}(q)$ is now equal to the right-hand side of Equation A.4].

Step 3:
Pre-multiplying $\bar{N}(q)$ by the r.s.p.m. of order k_i

$$N^{(i)}(q) = U^{(i)}(q) \bar{N}(q) \quad (A.6)$$

this multiplication shifts the matrix of coefficients of $\bar{N}(q)$, $\Lambda^{(i)}$, upwards by k_i rows of zeros. Update the matrix

$$D^{(i)}(q) = S^{(i)}(q) D^{(i-1)}(q) \quad (A.7)$$

This ends the ith iteration. Combining Equations A.4 to A.7, the ith iteration of the algorithm results in

$$N^{(i)}(q) = U^{(i)}(q) Q^{(i)} N^{(i-1)}(q) = S^{(i)}(q) N^{(i-1)}(q) = D^{(i)}(q) N(q)$$

where $S^{(i)}(q)$ and $D^{(i)}(q)$ are defined by Equations A.3 and A.7.
The final iteration ($t = i - 1$) yields

$$N^{(t)}(q) = D(q) N(q) \quad (A.8)$$

where $D(q) = D^{(t)}(q)$ is the unitary interactor matrix.

APPENDIX B
EXAMPLES OF THE DIAGONAL/GENERAL INTERACTOR MATRICES

The diagonal interactor matrix is relatively easy to obtain. For processes with diagonal interactor matrices, the smallest delay in each row is associated with the diagonal element of the matrix, i.e., each element, d_i, of the diagonal interactor matrix, $D = diag\{q^{d_1}, \cdots, q^{d_n}\}$, is actually the minimum delay in the ith row of the transfer function matrix. In other words, the diagonal interactor matrix depends solely on the minimum delay of each row of the transfer function matrix. Most interactor matrices of the *actual* multivariable process are either diagonal or general matrices (Goodwin and Sin 1984; Walgama 1986; Wolovich and Elliott 1983). The non-diagonal interactor matrix occurs when certain linear dependencies exist among the rows of the transfer function matrix (as $q^{-1} \to 0$) after the minimum delay of each row is factored out. For example, consider a 2×2 process

$$T = \begin{bmatrix} \frac{0.5q^{-2}}{1-0.7q^{-1}} & \frac{q^{-4}(1.4-0.9q^{-1})}{1-0.3q^{-1}} \\ \frac{1.23q^{-3}}{1-0.5q^{-1}} & \frac{4.7q^{-3}}{1-0.7q^{-1}} \end{bmatrix} \quad (B.1)$$

The minimum time delay of the first row is q^{-2} and the second row q^{-3}. After factoring out these minimum time delays from each row (this is equivalent to pre-multiplying T by a diagonal matrix $diag(q^2, q^3)$), we have the transfer function matrix

$$\tilde{T} = \begin{bmatrix} \frac{0.5}{1-0.7q^{-1}} & \frac{q^{-2}(1.4-0.9q^{-1})}{1-0.3q^{-1}} \\ \frac{1.23}{1-0.5q^{-1}} & \frac{4.7}{1-0.7q^{-1}} \end{bmatrix}$$

Thus,

$$\lim_{q^{-1} \to 0} \tilde{T} = \begin{bmatrix} 0.5 & 0 \\ 1.23 & 4.7 \end{bmatrix}$$

which is of full rank. Therefore, the interactor matrix is a diagonal matrix, i.e., $D = diag(q^2, q^3)$, but if the element $T_{1,2}$ of the transfer function matrix in Equation B.1 happens to be $\frac{q^{-2}(1.91-0.9q^{-1})}{1-0.3q^{-1}}$, then using the same diagonal factorization yields

$$\lim_{q^{-1} \to 0} \tilde{T} = \begin{bmatrix} 0.5 & 1.91 \\ 1.23 & 4.7 \end{bmatrix}$$

which is rank deficient, and a non-diagonal interactor matrix is then expected. In real processes, the exact linear dependency as in this illustration rarely

B. Examples of the diagonal/general interactor matrices

occurs. Another special case happens when the time delays associated with a particular input are larger than delays associated with other inputs. For example, if the transfer function of Equation B.1 is changed to

$$T = \begin{bmatrix} \frac{0.5q^{-2}}{1-0.7q^{-1}} & \frac{(1.4-0.9q^{-1})q^{-4}}{1-0.3q^{-1}} \\ \frac{1.23q^{-3}}{1-0.5q^{-1}} & \frac{4.7q^{-5}}{1-0.7q^{-1}} \end{bmatrix}$$

then the delays associated with second input are larger than the delays associated with the first input. Thus, using only the diagonal factorization will yield

$$\lim_{q^{-1} \to 0} \tilde{T} = \begin{bmatrix} 0.5 & 0 \\ 1.23 & 0 \end{bmatrix}$$

which is rank deficient so a non-diagonal interactor matrix is therefore expected. The existence of a general (non-diagonal) interactor matrix is usually due to this latter case.

In many multivariable processes under multiloop control, it is assumed implicitly that the input-output paring is such that the diagonal elements have smaller delays. This leads to the observation that the occurrence of a diagonal interactor matrix is not rare. More generally, even if a multivariable process has the minimum-delay pairing structure but is not paired in such a way in the actual multiloop design, the interactor matrix is still diagonal.

APPENDIX C
SPECTRAL TECHNIQUES IN PERFORMANCE ASSESSMENT

C.1 Introduction

It is a well known fact that mathematical equivalence of time series exists between its time domain expressions and its frequency domain expressions. Even though they are mathematically equivalent, we may sometimes extract more information by performing spectral analysis on time series data. The most common benchmark used in performance assessment has been the minimum variance benchmark. The minimum variance can be estimated from routine closed loop data with *a priori* knowledge of the time delay of the process. By comparing the autocorrelation or impulse response of this minimum variance output with the autocorrelation or impulse response of the actual output, one can determine how close the current controller is to the minimum variance condition. The same analysis can be done in the frequency domain where the minimum variance spectrum is compared to the spectrum of the actual output. It has been shown that this could give an indication of the frequency range where the output deviates from the minimum variance condition and of whether a controller is under- or overtuned(Desborough and Harris 1993). Spectral analysis techniques can play an important role in identifying some nonlinearities in the process(Thornhill, Oettinger, and Fedenczuk 1998). It is essential, during performance assessment, to separate problems related to a controller from actuator related problems. The spectrum analysis is applied to performance assessment of univariate processes in this appendix. The results are illustrated through two simulation examples.

C.2 Discrete Fourier Transform

The first step involved in spectral analysis is to transform the time domain data into frequency domain data. The data we deal with are usually discrete in nature. The discrete Fourier transform converts this discrete time domain data into frequency domain data. Consider time domain data given by y_t, t=1,2,...N where N is the total number of data points. Then the Fourier transform of this data is given by the following expression:

$$Y_N(w) = \frac{1}{\sqrt{N}} \sum_{t=1}^{N} y_t e^{-iwt} \tag{C.1}$$

These values are obtained at $w = 2\pi k/N$, k=1,....N.

The original series y_t can be recovered by the inverse Fourier transform given by

$$y_t = \frac{1}{\sqrt{N}} \sum_{k=1}^{N} Y_N(\frac{2\pi k}{N}) e^{i2\pi kt/N} \tag{C.2}$$

From the above relations we can conclude that

$$Y_N(w + 2\pi) = Y_N(w) \tag{C.3}$$

Therefore, the function $Y_N(w)$ is uniquely defined in the interval $[0, \pi]$ (Ljung 1998). The function $Y_N(w)$ is usually defined in the interval $-\pi \leq w \leq \pi$ so Equation C.2 is modified to

$$y_t = \frac{1}{\sqrt{N}} \sum_{k=-N/2+1}^{N/2} Y_N(\frac{2\pi k}{N}) e^{i2\pi kt/N} \tag{C.4}$$

The quantity $|Y(w)|^2$ represents the contribution of this frequency to the total energy of the signal y_t. This value is also known as the periodogram of the signal y_t. A useful relationship between the time domain and frequency domain expressions is obtained by Parseval's theorem given by

$$\sum \left| Y_N(\frac{2\pi}{N}k) \right|^2 = \sum y_t^2 \tag{C.5}$$

C.3 Power Spectrum

The autocovariance function of the signal y_t is given by

$$\gamma(\tau) = E[(y_t - \mu)(y(t+\tau) - \mu)] \tag{C.6}$$

where μ is the mean of the signal. The power spectrum of y_t is determined by

$$\Phi_y(w) = \sum_{\tau=-\infty}^{\infty} \gamma_y(\tau) e^{-iw\tau} \tag{C.7}$$

Also from the inverse Fourier transform, we obtain

$$\gamma(\tau) = \frac{1}{2\pi} \int_{-\pi}^{\pi} \Phi_y(w) e^{iw\tau} dw \tag{C.8}$$

Now if we assume that the signal y_t is related to a_t by a stable linear transfer function of the form

$$y_t = G_y(q^{-1})a_t$$

where the input a_t is a white noise sequence with mean zero and constant variance σ_a^2, then the following relationship holds

$$\Phi_y(w) = |G_y(e^{iw})|^2 \sigma_a^2 \qquad (C.9)$$

$G_y(q^{-1})$ is a time series model like a moving average model (MA), or, in general, an autoregressive moving average model (ARMA). A general stationary ARMA(p,q) model can also be represented as

$$\phi_p(q^{-1})y_t = \theta_q(q^{-1})a_t \qquad (C.10)$$

This can be transferred into an infinite order moving average model of the form

$$y_t = \sum_{j=0}^{\infty} f_j a_{t-j} = G_y(q^{-1})a_t \qquad (C.11)$$

where

$$G_y(q^{-1}) = \frac{\theta_q(q^{-1})}{\phi_p(q^{-1})} \qquad (C.12)$$

Therefore, one can estimate the spectrum of a signal through time series analysis.

C.4 Controller tuning

Let routine closed loop data y_t be represented as an infinite moving average series

$$y_t = [f_0 + f_1 q^{-1} + ... + f_{d-1} q^{-(d-1)} + ...]a_t = G_y(q^{-1})a_t \qquad (C.13)$$

Given process time delay as d sample intervals, the minimum variance output consists of the first d terms of the above expression; therefore, the minimum variance output can be written as

$$y_{t,mv} = [f_0 + f_1 q^{-1} + ... + f_{d-1} q^{-(d-1)}]a_t = G_{y,mv}(q^{-1})a_t \qquad (C.14)$$

The spectra of the above moving average model can be written as

$$\Phi_y(w) = |G_y(e^{iw})|^2 \sigma_a^2 \qquad (C.15)$$

and

$$\Phi_{y,mv}(w) = |G_{y,mv}(e^{iw})|^2 \sigma_a^2 \qquad (C.16)$$

The variance of the output y_t is obtained through

$$\sigma_y^2 = \frac{1}{2\pi} \int_{-\pi}^{\pi} \Phi_y(w) dw \qquad (C.17)$$

Since
$$\sigma_{y,mv}^2 \leq \sigma_y^2 \tag{C.18}$$
the following expression in frequency domain follows
$$\int \underbrace{[\Phi_{y,mv}(w) - \Phi_y(w)]}_{\delta(w)} dw \leq 0 \tag{C.19}$$

where $\delta(w)$ denotes the difference between the actual output spectrum and the minimum variance spectrum. Equation C.19 represents the area under the spectral plot. This means that the area under the minimum variance spectrum is always lesser than the area under the output spectrum. If, at some frequency, the output spectrum has a lower variance than minimum variance, it implies that the output variance at other frequencies must be greater than that of the minimum variance. For example, a typical overtuned controller may result in an underdamped closed-loop response and oscillatory behavior in the output response. The oscillations may be detected in the spectrum as peaks. Unfavourable response characteristics associated with disturbances can also result in periodicities which can be diagnosed by spectral analysis. It is, therefore, important to separate the periodicities due to controller tuning and those due to unfavourable disturbances. In the case of measured disturbances, the autospectrum of these disturbances may indicate whether any periodicities exist so that corrective action can be taken at the source and not by retuning the feedback controller. Another source of cycling in the closed-loop response may be due to valve stiction, hysteresis *etc.*, Bilkaowski (1993) have reported that poor valve design and maintenance contribute significantly to the output variance.

C.5 Simulation example

The simulation example deals with a process represented by
$$G(q^{-1}) = \frac{.33z^{-4}}{1 - 0.67q^{-1}} \tag{C.20}$$
It is controlled by a Dahlin controller given by
$$G_c(q^{-1}) = \frac{0.5[.7 - 0.47q^{-1}]}{0.33 - .1q^{-1} - .23z^{-4}} \tag{C.21}$$

The sampling time is one minute. The minimum variance output and its spectrum are estimated from the routine closed-loop data. Figure C.1 shows the spectrum of the minimum variance output together with the spectrum of the actual output. Three different controller settings are studied, and results are shown in the figure. The first controller setting is given by

$$G_c(q^{-1}) = \frac{0.3[.7 - 0.47q^{-1}]}{0.28 - .1q^{-1} - .23z^{-4}} \quad (C.22)$$

It can be seen that the actual spectrum deviates from the minimum variance spectrum. A more aggressive controller given by

$$G_c(q^{-1}) = \frac{0.8 * [.7 - 0.47q^{-1}]}{0.33 - .1q^{-1} - .23z^{-4}} \quad (C.23)$$

is then employed. This controller setting makes the spectrum of the output shift below that of the minimum variance spectrum. To make the spectrum closer to the minimum variance spectrum, a controller slightly detuned from the one above is implemented. This controller setting is close to the minimum variance condition. It is worth mentioning that we do not necessarily want to reach minimum variance condition because it may pose problems like large input moves *etc.*; therefore, this setting that is close to the minimum variance condition may be regarded as a satisfactory controller setting.

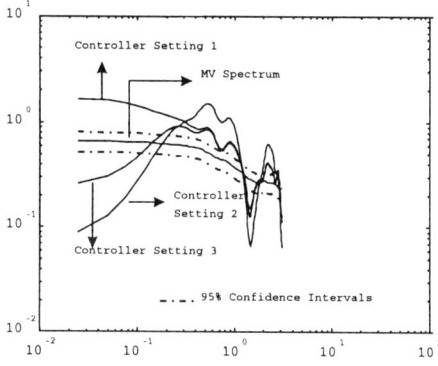

Fig. C.1. Comparison of performance through spectrum analysis.

REFERENCES

Aoki, M. (1987). *State Space Modeling of Time Series*. Berlin Heidelberg: Springer-Verlag.

Astrom, K. (1967). Computer control of a paper machine - an application of linear stochastic control theory. *IBM J. 11*, 389–405.

Astrom, K. (1970). *Introduction to Stochastic Control Theory*. New York: Academic Press.

Astrom, K. and B. Wittenmark (1990). *Computer-Controlled Systems, Theory and Design* (Second ed.). Prentice-Hall.

Basseville, M. (1998) On-board component fault detection and isolation using the statistical local approach. *Automatica 34*(11), 1391–1415.

Basseville, M. and I. Nikiforov (1993). *Detection of Abrupt Changes*. Englewood Cliffs, New Jersey: Prentice Hall.

Bialkowski, W. (1993). Dreams versus reality: A view from both sides of the gap. *Pulp and Paper Canada 94*(11), 19–27.

Bitmead, R., M. Gevers, and V. Wertz (1990). *Adaptive Optimal Control*. Prentice Hall.

Bittanti, S., P. Colaneri, and M. Mongiovi (1994, December). The spectral interactor matrix for the singular Riccati equation. In *Proceedings of the 33rd CDC*, pp. 2165–2169.

Borison, U. (1979). Self-tuning regulators for a class of multivariable systems. *Automatica 15*, 209–215.

Box, G. and G. Jenkins (1976). *Time Series Analysis Forecasting and Control*. Holden-Day.

Box, G. and J. MacGregor (1974). the analysis of closed-loop dynamic stochastic systems. *Technometrics 18*, 371–380.

Box, G. and J. MacGregor (1976). Parameter estimation with closed-loop operating data. *Technometrics 18*, 371 – 380.

Boyd, S. and C. Barratt (1991). *Linear Control Design*. Prentice Hall, Eaglewood Cliffs, New Jersy.

Caines, P. and C. Chan (1975). Feedback between stationary stochastic processes. *IEEE Trans AC 20*, 498–508.

Chu, C. (1985). H_∞-*Optimization and Robust Multivariable Control*. Ph. D. thesis, University of Minnesota, Minneapolis, MN.

Clarke, D., C. Mohtadi, and P. Tuffs (1987). Generalized predictive control - part 1 and 2. *Automatica 23*(2), 137–160.

Dahleh, M. and I. Diaz-Bobillo (1995). *Control of Uncertain Systems–a Linear Programming Approach*. Englewood Cliffs, New Jersey: Prentice Hall.

Defalque, B., M. Gevers, and M. Installe (1976). Combined identification of the input-output and noise dynamics of a closed-loop controlled linear system. *Int. J. Control 24*, 345–360.

References

Desborough, L. and T. Harris (1992). Performance assessment measure for univariate feedback control. *Can. J. Chem. Eng. 70*, 1186–1197.

Desborough, L. and T. Harris (1993). Performance assessment measures for univariate feedforward/feedback control. *Can. J. Chem. Eng. 71*, 605–616.

Desborough, L. and T. Harris (1994). Control performance assessment. *Pulp and Paper Canada 94*(11), 441.

DeVRIES, W. and S. Wu (1978, August). Evaluation of process control effectiveness and diagnosis of variation in paper basis weight via multivariate time series analysis. *IEEE Trans. on AC AC-23, No 4*.

Dugard, L., G. Goodwin, and X. Xianya (1984). The role of the interactor matrix in multivariable stochastic adaptive control. *Automatica 20*(5), 701–709.

Eriksson, P. and A. Isaksson (1994). Some aspects of control performance monitoring. In *Proc. 3rd IEEE Conf.. Control Applications*, Glasgow, Scotland, pp. 1029–1034.

Forssell, U. and L. Ljung (1999). Closed-loop identification revisited. *Automatica 35*(7), 1215–1241.

Garcia, C., D. Prett, and M. Morari (1989). Model predictive control: Theory and practice - a survey. *Automatica 25*(3), 335–348.

Gevers, M. (1993, June). Towards a joint design of identification and control. In *2nd European Control Conference*, Holland.

Gevers, M. and L. Ljung (1986). Optimal experiment designs with respect to the intended model application. *Automatica 22 No 5*, 543 – 554.

Golub, G. and C. V. Loan (1989). *Matrix Computations* (Second ed.). The Johns Hopkins University Press.

Goodwin, G. and R. Payne (1977). *Dynamic System Identification: Experiment Design and Data Analysis*. New York: Academic Press.

Goodwin, G. and K. Sin (1984). *Adaptive Filtering Prediction and Control*. Englewood Cliffs: Prentice-Hall.

Guo, L. and L. Ljung (1994). The role of model validation for assessing the size of the unmodelled dynamics. In *Proceedings of CDC*.

Gustavsson, I., L. Ljung, and T. Soderstrom (1978). Identification of processes in closed loop — identifiability and accuracy aspects. *Identification and System Parameter Estimation*, 41–77.

Hagglund, T. (1995). A control-loop performance monitor. *Control Engineering Practice 5*(11), 1543.

Hakvoort, R., R. Schrama, and P. Van den Hof (1994). Approximate identification with closed-loop performance criterion and application to LQG feedback design. *Automatica 30*(4), 679–690.

Harris, T. (1985). A comparative study of model based control strategies. In *Proceedings of ACC*.

Harris, T. (1989). Assessment of closed loop performance. *Can. J. Chem. Eng. 67*, 856–861.

Harris, T., F. Boudreau, and J. MacGregor (1995). Performance assessment of multivariable feedback controllers. In *1995 AIChE Annual Meeting (also in Personal Communication)*, Miami Beach, FL.

Harris, T., F. Boudreau, and J. MacGregor (1996). Performance assessment of multivariable feedback controllers. *Automatica 32*(11), 1505–1518.

Harris, T. and J. MacGregor (1987). Design of multivariable linear quadratic controllers using transfer functions. *AIChE J. 33*, 1481–1495.

Harris, T., C. Seppala, and L. Desborough (1999). A review of performance monitoring and assessment techniques for univariate and multivariate control systems. *J. of Process Control 9*, 1–17.

Harris, T., C. Seppala, P. Jofriet, and B. Surgenor (1996). Plant-wide feedback control performance assessment using an expert system framework. *Control Engineering Practice 4*, 1297.

Hjalmarsson, H., M. Gevers, F. Bruyne, and J. Leblond (1994, December). Identification for control: Closing the loop gives more accurate controllers. In *Proceedings of 1994 CDC*, Lake Buena Vista, FL, pp. 4150–4155.

Huang, B. (1999a, May 9-12). Process and control loop performance monitoring through detection of abrupt changes of closed-loop models. In *Proc. of 1999 IEEE Canadian Conference on Electrical & Computer Engineering*, Edmonton, Canada, pp. 1559–1564.

Huang, B. (1999b, July). System identification based on last principal components analysis, to appear. In *Proceedings of IFAC'99 World Congress*, Beijing, China.

Huang, B., S. Shah, and H. Fujii (1996, July). Identification of the time delay/interactor matrix for MIMO systems using closed-loop data. In *Proc. 13th IFAC World Congress*, Volume M, San Francisco, pp. 355–360.

Huang, B., S. Shah, and K. Kwok (1995, June). On-line control performance monitoring of MIMO processes. In *Proc. American Control Conference*, Seattle, Washington, pp. 1250–1254. American Control Conference.

Huang, B., S. Shah, and K. Kwok (1996, July). How good is your controller? application of control loop performance assessment techniques to MIMO processes. In *Proc. 13th IFAC World Congress*, Volume M, San Francisco, pp. 229–234.

Huang, B., S. Shah, and R. Miller (1999). Feedforward plus feedback controller performance assessment of MIMO systems. *To appear in IEEE Trans. on Control System Technology*.

Huang, B. and E. Tamayo (1998). Advances in control system performance monitoring via model validation. In *Proc. of 1998 International Association of Science and Technology Control Applications Conference*.

Huang, B., X. Zhao, E. Tamayo, and A. Hanafi (1999, May 9-12). Fault diagnosis of an industrial CGO coker model predictive control system. In *Proc. of 1999 IEEE Canadian Conference on Electrical & Computer Engineering (to appear)*, Edmonton, Canada, pp. 960–965.

Jofriet, P. and W. Bialkowski (1996). The key to on-line monitoring of process variability and control loop performance. In *Proc. Control Systems'96*, pp. 187.

Jofriet, P., C. Seppala, M. Harvey, B. Surgenor, and T. Harris (1996). An expert system for control loop performance. In *Pulp and Paper Canada*, pp. 207.

Kadali, R., B. Huang, and E. Tamayo (1999). A case study on performance analysis and trouble shooting of an industrial model predictive control system. *To appear in proc. of 1999 American Control Conference*.

Kendra, S. and A. Cinar (1997). Controller performance assessment by frequency domain techniques. *Journal of Process Control 7*(3), 181–194.

Kesavan, P. and J. Lee (1997). Diagnostic tools for multivariable model-based control systems. *Ind. Eng. Chem. Res. 36*, 2725–2738.

Keviczky, L. and J. Hetthessy (1977, January). Self-tuning minimum variance control of MIMO discrete time systems. *Control Theory and Applications 5*(1).

Kosut, R., G. Goodwin, and M. Polis (1992). Special issue on system identification for robust control design. *IEEE Trans Automatica Control 37*.

Kozub, D. Controller performance monitoring and diagnosis: Experience and challenges. *AIChE Symposium Series 93*(316), 83.

Kozub, D. and C. Garcia (1993, Nov. 9). Monitoring and diagnosis of automated controllers in the chemical process industries. St. Louis, MO. AIChE Annual Meeting.

Kwakernaak, H. and R. Sivan (1972). *Linear Optimal Control System*. John Wiley & Sons.

Ljung, L. (1994). System identification in a MIC perspective. *Modeling, Identification and Control* 15(3), 153 – 159.

Ljung, L. (1998). *System Identification* (2nd ed.). Prentice-Hall.

Lynch, C. and G. Dumont (1993, Sept.13-16). Closed loop performance monitoring. In *Proc. 2nd IEEE Conf. Control Applications*, Vancouver, B.C., pp. 835 – 840.

Lynch, C. and G. Dumont (1996). Control loop performance monitoring. *IEEE Trans. Control Sys. Tech* 4(2), 185–192.

MacGregor, J. (1976). Optimal choice of the sampling interval for discrete process control. *Technometrics* 18(2), 151–160.

MacGregor, J. and D. Fogal (1995). Closed-loop identification: The role of the noise model and prefilters. *Journal of process control* 5(3), 163–171.

MacGregor, J., T. Harris, and J. Wright (1984). Duality between the control of processes subject to randomly occuring deterministic disturbances and ARIMA stochastic disturbances. *Technometrics 26*.

Miao, T. and D. Seborg (1995, June). A monitoring strategy for flow and pressure control loops. In *The 2nd Asia-Pacific Conf. on Control and Measurement*, Wuhan-Chongqing, China.

Miller, R. (1995). Stochastic predictive control. Master's thesis, University of Alberta.

Miller, R. and B. Huang (1997). Perspectives on multivariate feedforward/feedback controller performance measures for process diagnosis. In *Proceedings of AD-CHEM*, Banff, Canada, pp. 435.

Mohtadi, C. (1988). On the role of prefiltering in parameter estimation and control. In S. S.L. and G. Dumont (Eds.), *Adaptive Control Strategies for Industrial Use*, pp. 121 – 138. Springer-Verlag.

Morari, M. and E. Zafiriou (1989). *Robust Process Control*. Prentice Hall.

Mutoh, Y. and R. Ortega (1993). Interactor structure estimation for adaptive control of discrete-time multivariable nondecouplable systems. *Automatica* 29(3), 635–647.

Newcombe, D. (1991). Roll headboxes and their approach flow. *Pulp and Paper Manufacture 7*, 97–116.

Nordstrom, B. and B. Norman (1994). Influence of headbox nozzle contraction ratio on sheet formation and anisotropy. In *Proceedings of the 1994 Engineering Conference*, TAPPI, pp. 225–228.

Owen, J., D. Read, H. Blekkenhorst, and A. Roche (1996). A mill prototype for automatic monitoring of control loop performance. In *Proc. Control Systems'96*, pp. 171.

Paige, C. (1981). Properties of numerical algorithms related to computing controllability. *IEEE Trans. Auto. Control AC-26*, 130–138.

Paplinski, A. and M. Rogozinski (1990). Right nilpotent interactor matrix and its application to multivariable stochastic control. In *Proceedings of ACC*, Volume 1, pp. 494–495.

Peng, Y. and M. Kinnaert (1992, May). Explicit solution to the singular LQ regulation problem. *IEEE Trans AC 37*, 633–636.

Phadke, M. and S. Wu (1974). Identification of multi-input-multi-output transfer function and noise model of a blast furnace from closed-loop data. *IEEE Trans AC 19*, 944–951.

Pierce, D. (1975). Forecasting in dynamic models with stochastic regressors. *Journal of Econometrics 3*, 349–374.

Qin, S. (1998). Control performance monitoring - a review and assessment. *Computers and Chemical Engineering 23*, 173–186.

Reinsel, G. (1993). *Elements of Multivariate Time Series Analysis*. New York: Springer-Verlag.

Rhinehart, R. (1995). A watch dog for controller performance monitoring. In *Proceedings of the 1995 American Control Conference*, Seattle, Washington, U.S.A., pp. 2239 – 2240.

Rice, J. (1972, November 21). U.S. patent 3,703,436. Noyes Data Corporation. Park Ridge New Jersey.

Rivera, D., J. Pollard, and C. Garcia (1992, July). Control-relevant prefiltering: A systematic design approach and case study. *IEEE Tran AC 37*(7), 964–974.

Rogozinski, M., A. Paplinski, and M. Gibbard (1987, March). An algorithm for calculation of a nilpotent interactor matrix for linear multivariable systems. *IEEE Trans. AC 32*(3), 234–237.

Schrama, R. (1991). An open-loop solution to the approximate closed-loop approximation problem. In *Proceedings of IFAC identification and system parameter estimation*, Budapest, Hungary, pp. 761–766.

Schrama, R. (1992). *Approximate Identification and Control Design*. Ph. D. thesis, Delft University of Technology.

Shah, S., C. Mohtadi, and D. Clarke (1987). Multivariable adaptive control without a prior knowledge of the delay matrix. *Systems and Control Letters 9*, 295–306.

Shook, D., C. Mohtadi, and S. Shah (1992, July). A control-relevant identification strategy for GPC. *IEEE Trans. AC 37 NO. 7*.

Soderstrom, T. and P. Stoica (1989). *System Identification*. UK: Prentice Hall International.

Sripada, N. (1988). *Multi-step Adaptive Predictive Control with Disturbance Modeling*. Ph. D. thesis, Department of Chemical Engineering, University of Alberta, Edmonton, Alberta, Canada.

Stanfelj, N., T. Marlin, and J. MacGregor (1993). Monitoring and diagnosing process control performance: the single-loop case. *Ind. Eng. Chem. Res. 32*, 301–314.

Sternad, M. and T. Soderstrom (1988). LQG-optimal feedforward regulators. *Automatica 24*(4), 557–561.

Thornhill, N., M. Oettinger, and P. Fedenczuk (1998). Refinery-wide control loop performance assessment. *to appear in Journal of Process Control*.

Thornhill, N., R. Sadowski, R. Davis, J. Fedenczuk, P. Knight, M. Prichard, and D. Rothenberg (1996). Practical experiences in refinery control loop performance assessment. In *UKACC'96*.

Tiao, G. and G. Box (1981). Modeling multiple time series with application. *JASA 76*, 802–816.

Tsiligiannis, C. and S. Svoronos (1988). Dynamic interactors in multivariable process control-1, the general time delay case. *Chemical Engineering Science 43*(2), 339–347.

Tsiligiannis, C. and S. Svoronos (1989). Dynamic interactors in multivariable process control-2, time delays and zeros outside the unit circle. *Chemical Engineering Science 44*(9), 2041–2047.

Tyler, M. and M. Morari (1995). Performance assessment for unstable and nonminimum-phase systems. Technical Report IfA-Report No 95-03, California Institute of Technology.

Tyler, M. and M. Morari (1996). Performance monitoring of control systems using likelihood methods. *Automatica 32*, 1145–1162.

Van den Hof, P. and R. Schrama (1993). An indirect method for transfer function estimation from closed loop data. *Automatica 29*(6), 1523–1527.

Van den Hof, P. and R. Schrama (1995). Identification and control—closed-loop issues. *Automatica 31*(12), 1751–1770.

Vishnubhotla, A., S. Shah, and B. Huang (1997). Feedback and feedforward performance analysis of the shell industrial closed-loop data set. In *Proceedings of ADCHEM*, Banff, Canada, pp. 295–300.

Walgama, K. (1986). Multivariable adaptive predictive control for stochastic systems with time delays. Master's thesis, University of Alberta.

Wiener, N. (1949). *Extrapolation, Interpolation & Smoothing of Stationary Time Series*. Cambridge, Mass.: MIT Press.

Wolovich, W. and H. Elliott (1983). Discrete models for linear multivariable systems. *Int. J. Control 38*(2), 337–357.

Wolovich, W. and P. Falb (1976). Invariants and canonical forms under dynamic compensation. *SIAM J. Control 14*, 996–1008.

Youla, D. and J. Bongiorno (1985). A feedback theory of two-degree-of-freedom optimal Wiener-Hopf design. *IEEE on AC 30*.

Zang, Z., R. Bitmead, and M. Gevers (1995). Iterative weighted least-squares identification and weighted LQG control design. *Automatica 31*(11), 1577–1594.

Zhang, Q., M. Basseville, and A. Benveniste (1998). Fault detection and isolation in nonlinear dynamic systems: A combined input-output and local approach. *Automatica 34*(11), 1359–1373.

Zhu, Y. and T. Backe (1993). *Identification of Multivariable Industrial Processes*. Springer-Verlag.

INDEX

H_2 norm, 127

Absolute lower bound, 9, 68, 87
Absorption process, 94
Admissible minimum variance control, 111, 112, 116
All-pass factor, 26, 134
AR model, 14
ARMA model, 14
Asymptotic variance and bias error, 185
Asymptotic variance of estimate, 183
Autocorrelation function, 216, 223

Backshift operator, 10
Bias error, 181, 184
Bilinear transformation, 108

Capacitance drum, 230
Cascade control, 15
Catalytic reaction, 15
Change of disturbance dynamics, 175
Closed-loop, 44
Closed-loop identification, 183
Closed-loop potential or CLP, 12
Closed-loop transfer function, 13
Composition controller, 225
Control loop performance assessment, 9
Controller sampling frequency, 70
Controller transfer function, 10
Correlation analysis, 45, 60
Cross-correlation, 12

Dahlin-controller, 128
Data sampling frequency, 70
Data segmentation, 169
DCS, 213
Deadbeat control, 19, 26
Delay-free, 10
Delay-free transfer function matrix, 21
Desired impulse response, 139

Desired transfer function matrix, 144
Detection of abrupt changes, 169
Detection of small changes, 169
Determination of numerical rank, 46
Determination of the order of the interactor matrix, 38
Deterministic disturbance, 123
Diagonal interactor matrix, 20, 65
Different ordering of the output variables, 25
Diophantine identity, 10, 33, 58, 112, 115
Direct closed-loop identification, 183
Discrete Fourier transform, 241
Distillation column, 51
Disturbance transfer function, 10

Early warning of process change (fault), 171
Equivalence of weighted singular LQ and minimum variance control, 32
Estimated innovations sequence, 13
Estimated white noise, 14
Estimation of the admissible minimum variance control variance, 113
Estimation of the interactor matrix under closed-loop conditions, 44
Estimation of the Markov parameters, 38
Estimation of the minimum variance term, 59, 67, 84
Estimation of the unitary interactor matrix, 38
Example of interactor matrices, 239
Example of interactor matrix estimation, 47
Excessive control action, 87
Existence of the generalized unitary interactor matrix, 108
Existence of the weighted unitary interactor matrix, 30, 31

Explicit optimal control law, 26
Extended horizon performance index, 214

Factorization of the unitary interactor matrix, 41, 237
False alarm rate, 172
FCOR algorithm, 13, 62, 70, 81, 90
Feedback control invariant, 9, 45, 57, 83, 114
Feedforward & feedback control, 99
Feedforward control, 99
Finite zero, 20
First-level benchmark, 87
First-level performance assessment, 133
Frequency response, 223
Fundamental limitation on performance, 19

Gain updating, 169
General interactor matrix, 20, 81, 90
Generalized likelihood ratio test, 172
Generalized predictive control or GPC, 155
Generalized unitary interactor matrix, 108, 146

Headbox, 71
Higher-level performance assessment, 133
Historians, 213

Ideal factorization, 26
IMC, 115, 137, 143, 148
Impulse response, 214, 216
Impulse response matrix, 24
Infinite moving-average, 11
Infinite zero, 19, 40
Inner loop, 17
Inner-outer factorization, 26, 42, 109
Innovation sequence, 14
Interactor filtered output, 26
Interactor matrix, 19, 20, 39
Invariance of the interactor matrix under feedback control, 44
Inverse Fourier transform, 242
Inverse sensitivity function, 188
ISE, 23

Joint identification and control, 186

Kalman filter, 14, 155

Local approach, 169

Lower left triangular interactor matrix, 20, 21
LQ control, 24
LQ objective function, 28, 82, 109
LQG benchmark, 157

Markov parameter matrix, 24, 39, 41, 42, 46, 81, 103
Mean square error, 90
Measurable disturbance, 100
MIMO closed-loop identification, 195
MIMO model, 19
Minimum H_2-norm, 114
Minimum feedforward and feedback control variance term, 102
Minimum SSE control, 124
Minimum variance control, 11, 19, 24, 26, 27, 57
Minimum variance control law, 59
Minimum variance control via spectral factorization, 32
Minimum variance feedforward and feedback control, 101
Minimum variance feedforward and feedback term, 104
Minimum variance spectrum, 241
Minimum variance term, 11, 58, 59, 83–85
Model predictive control, 155, 169
Model validation, 169
Monte-Carlo simulation, 176
Moving average, 12, 58, 84
Multiloop PID controller, 78
Multivariable feedback controller, 19
Multivariable minimum variance control, 24
Multivariable process, 19
Multivariate, 19
Multivariate performance index, 61

Nilpotent interactor matrix, 20, 41
Non-minimum phase zero, 68, 108, 146, 192
Non-square transfer function matrix, 85, 93
Non-uniqueness of the interactor matrix, 25
Non-uniqueness of the unitary interactor matrix, 30
Normalized multivariate impulse response, 224
Normalized residual, 171
Number of infinite zeros, 20

Numerical example of multivariable minimum variance control, 34

Optimal singular LQ control, 24
Order of the interactor matrix, 20, 40–42
Outer loop, 17
Output error method, 185
Output-order dependent, 28
Output-order invariant, 29

Parseval's theorem, 28
Performance index, 12, 13, 60, 61, 69, 90, 127
Performance limitation, 19
Periodogram, 242
Plant transfer function, 10
Poor controller, 133
Power spectrum, 242
Power spectrum density, 28
Pre-whitening, 13
Prediction error, 181
Prediction error method or PEM, 181
Primary residual, 171
Process offset, 61, 90
Process quality control, 169
Process re-engineering, 10
Proper transfer function, 11
Pulp/paper process, 71

QR decomposition, 24

Random setpoint, 75
Real-time SIMULINK, 206
Realistic benchmark, 133
Regression analysis, 173
Regulatory control, 10, 122
Residual test, 217
Residuals, 14
Right matrix fraction, 237
Right unitary interactor matrix, 26
Robustness, 87
Routine operating data, 9
Row shift polynomial matrix, 237

Scaling invariant, 29
Second level performance assessment, 107
Second-level benchmark, 88
Sensitivity function, 183, 185, 188
Setpoint tracking, 122
Settling time benchmark, 136
Shaping filter, 191
Signal to noise ratio or SNR, 183

Simple interactor matrix, 20, 57, 85
Simulation of SISO performance assessment, 15
Singular LQ control, 24, 32, 82
Singular value decomposition, 40, 46
SISO process, 10
Solutions for the weighted unitary interactor matrix, 32
Spectral analysis, 217, 241
Spectral factorization of the interactor matrix, 90
Spectral interactor matrix, 26
Statistical test, 171
Stochastic disturbance, 123
Stochastic framework, 19

Time delay, 9
Time series analysis, 9
Time variant, 170
Total variance, 102
Tradeoff curve, 156, 157
Transfer function, 10
Triangular interactor matrix, 20, 28
Two-interacting-tank process, 159
Two-step closed-loop identification, 186
Two-step closed-loop identification algorithm, 191

Underestimation of the performance index, 71
Uniqueness of minimum variance control, 30
Unitary interactor matrix, 20, 21, 23, 24, 41, 81, 85, 143
Univariate process, 10
Unmeasurable disturbance, 99
User-specified benchmark, 135, 138, 147

Variance error, 181

W-filtering method, 192
Weighted LQ objective, 83
Weighted minimum variance control, 86
Weighted singular LQ control, 32
Weighted unitary interactor matrix, 30, 32
White noise sequence, 9, 10
Whitening, 14
Wood-Barry distillation column, 222

Y-filtering method, 192

Z-transform, 10